AF693368

DEVELOPMENT DYNAMICS IN COMMAND AREAS OF MAJOR IRRIGATION PROJECT

DEVELOPMENT DYNAMICS IN COMMAND AREAS OF MAJOR IRRIGATION PROJECT

R. RAM MOHAN RAO
&
S. SIMHADRI

INTER-INDIA PUBLICATIONS
D-17, RAJA GARDEN EXTN. NEW DELHI-110 015 (INDIA)
PHONE: 5413145

Cataloging in Publication Data--DK

Ram Mohan Rao, R. (Rairao), 1943-

Development dynamics in command areas of major irrigation projects.

Bibliography: p.

Includes index.

1. Sriramsagar Command Area. 2. Irrigation farming --India--Andhra Pradesh. 3. Irrigation--India--Andhra Pradesh. I. Simhadri, S. (Somanaboina), 1955- joint author. II. Title.

ISBN 81-210-0236-2

First Published in India in 1989
© R. Ram Mohan Rao & S. Simhadri 1988
Rairao Ram Mohan **Rao** *(b. 1943-)*
Somanaboina **Simhadri** *(b. 1955-)*

Published in India by:
M.C. Mittal
INTER-INDIA PUBLICATIONS
D-17, RAJA GARDEN EXTN.,
NEW DELHI-110015 (INDIA)
PHONE : 5413145

Printed in India by

A.P. Offset
Panchsheel Garden, Naveen Shahadra
Delhi-110032

Preface

A preponderant feature of India's post-independence developmental planning is the construction of major irrigation projects. These gigantic, high investment irrigation projects were primarily aimed at dilating agricultural productivity potential levels by providing assured water supply to the crops in their command areas. An attempt has been made in the present study to analyse the process of metamorphosis that is underway in Sriramsagar Command Area in respect of population characteristics, agricultural system, settlement pattern, social aspects and levels of regional development. The volume is the outcome of a University Grants Commission sponsored study which spanned over a period of two and a half years. We would like to put on record our thanks to University Grants Commission.

The obligations both general and specific, that we have incurred in the preparation of this volume are numerous. The writers of any such work as this, draws upon a body of knowledge built up through the efforts of several researchers over many years. Our indebtedness to these scholars is acknowledged with pleasure. We are specially grateful to Prof. P.D. Mahadev of Mysore University with whom we had the opportunity of discussing the subject matter.

The authors would like to extend their sincere thanks to the members of the research staff. Mr. Satish Chandra, a statistician by training, was largely responsible for statistical analysis of the data and for the preparation of sample design. Mr. S. Rama Krishna who also worked with us for some months, assisted us very ably in research work. Our thanks are also due to Mr. Satyanarayana Reddy, Mr. Rajaram Rao for their help in conducting field studies. We are thankful to various private and governmental agencies who have been always obliging in complying with our requests for data and information.

The maps in this report were drawn by Mr. Ikramullah. We fully appreciate his cartographic expertise. The assistance provided by Mr. Padmanabha Rao in cartographic and data processing work is gratefully acknowledged. We are grateful to the Head, Department of Geography, Osmania University for providing the necessary infrastructure facilities in the Department. Although we should like to believe that the text of this report is free from mistakes, printing and other, we may not do so. We shall be obliged to scholars who will inform us of errors they may detect, so that we can correct the same in subsequent editions.

R. RAM MOHAN RAO
S. SIMHADRI

Contents

List of Tables

List of Figures

List of Appendices

ONE

Introduction

KRISHNA and Godavari are the two most important rivers of peninsular part of India. Since times immemorial, these rivers have been playing a crucial role in shaping the economy and culture of the region, south of Vindhyas in general and the present State of Andhra Pradesh in particular. Andhra Pradesh would not have been known as the 'Granary of the South' but for the passage of these rivers across the State. However, the benefits as well as the ill-effects of the rivers were confined, till recently, only to a small part of the course, namely, their deltaic parts. When India became independent, this anamolous situation was sought to be remedied by constructing dams across the rivers in the middle and upper parts of the valleys so that the waters can be provided to the parched lands and restricting flow of flood water in the lower reaches. As a follow up measure of this policy, two major dams and reservoirs, namely, Nagarjunasagar and Sriramsagar were built across the rivers Krishna and Godavari. Sriramsagar was hitherto known as Pochampad Project.

Nagarjunasagar project was commissioned in the sixties of the present century. It was built to store and supply water to irrigate an area of one million hectares falling in the districts of Guntur, Prakasam, Khammam, Krishna and Nalgonda.

The Pochampad Project, since renamed as Sriramsagar Project, was planned to be built across the River Godavari to irrigate nearly seven lakh hectares of land in the first and second phases. The Command area of this river project encompasses the districts of Adilabad, Karimnagar, Nizamabad, Warangal and Nalgonda and Khammam. The project is essentially an irrigation project involving building of a 140 feet high dam across Godavari at Pochampad and, the construction of two main canals, namely, Kakatiya canal and Saraswathi canal. Kakatiya canal which originates on the right side of the river flows through the districts of Karimnagar and Warangal during the first phase of its construction. The canal will ultimately reach Khammam through Nalgonda district if the second phase of its construction is taken up. The north canal named Saraswathi canal is proposed to be built for a total length of 200 km for an ayacut of 1.45 lakh acres in addition to stabilishing 0.30 lakh hectares under the Kadam project. The water was first released into Kakatiya canal in 1970 up to 36 km to irrigate 10,000 hectares. By 1978, water was released to benefit a total area of 1 lakh hectares. An irrigation potential of 2.2 lakh hectares was created by the end of the year 1985.

The present work was conceived with the objective of studying the socio-economic transformation that has taken place in the command area of Sriramsagar irrigation project. The socio-economic transformation may be observed in the changing land-use and cropping patterns, in demographic aspects, in agricultural system, in social characteristics of population, in the pattern of distribution of settlements and in levels of development.

The authors of this book had an opportunity to tour intensively in the left canal command area of Nagarjunasagar in 1978. Subsequently, a number of intensive field studies were conducted in the selected villages of the command area. It was found during the course of such studies that although, there was no doubt a lot of agricultural development in the region due to intensive cultivation prompted by the provision of assured irrigation facilities, the benefits of the project did not reach for whom they were meant, as revealed by an indepth study (Rao, R.R.M., 1978). Why did it happen? To find an answer to this question, it is necessary to know the type of agricultural system that was in existence in the area. The left canal command area of Nagarjunasagar project has traditionally been a drought prone, rain dependant-crop growing area. The local farmers were not used to irrigated intensive farming and lacked the necessary skills as well as the financial back-up that are so essential in such a system. The people from the neighbouring Andhra districts who have been familiar with intensive irrigation farming methods and who were affluent and dynamic enough to extend their farming activities to new areas, found new vistas of enterprise in the adjoining command area of Nagarjunasagar project. They purchased agricultural lands in the command area just before commissioning of the project from the local farmers at prices which superficially seemed to be good and higher than the existing prices of dry lands in the area. But the prices at which these lands were brought by Coastal Andhra people were not even a fraction of the prices that the same type of land in the delta region would sell. Once the waters were released to irrigate the lands in the command area, benefits started flowing in by the way of higher outputs etc. Who were the beneficiaries? Certainly not the majority of the local farmers for whose benefit the project was constructed. Andhra farmers were very successful in increasing the agricultural productivity of the land by making use of their skills in intensive farming techniques. They were affluent enough to be able to apply higher amounts of inputs in the form of fertilizers, better seeds, pesticides etc. and they could get commensurate benefits. Some of the local farmers who did not or could not sell their lands, are now faced with the task of adopting themselves to changed scenario. Some were successful; many were not so successful. They encountered the problems such as mobilising finances, acquiring skills etc.

The Andhra migrants not only brought in changes in the agricultural land because of their superior skills, but they also changed the pattern of settlement structure by implanting their native type of settlements in the command area. They built houses, and in some cases the entire settlements were modelled on Coastal Andhra Pattern. Marked transformation took place with regard to general lay-out of the house, grain storage facilities, type of cattle shed etc. Enquiries by the authors revealed that there was a change in the social structure, in social values, habits and in attitudes as well.

With a background of the information described above, it was thought that it would be worth-while to make detailed investigation of Sriramsagar command area. The book has been organised into seven chapters with the objective of discussing and analysing such aspects as physical setting, agricultural system, pattern and levels of regional development. It is perhaps necessary to mention here that the study, though investigative in approach, is mostly a base line study. It has been patterned in this manner since the entire area has not yet received the irrigation waters and even the areas which are irrigated, are yet to feel the full impact of the release of water on their economy. However, attempts have been made, wherever it was felt necessary, to compare the irrigated and unirrigated components. It is thought that base line study would be helpful in future when a subsequent follow up study is made after a time gap, so that the dynamics of the effects of irrigation development can be worked out and probed with a view to make generalisations.

Data Base

Population data have been collected, processed and analysed at micro-level *i.e.* village as the unit. The enormity of data collection work can be gauged from the fact that there are over seven hundred settlements spread over four districts and sixteen taluks in Sriramsagar command area. Collection, processing, cartographic representation and analysis of data, though consumed major chunk of the stipulated time frame, was worthwhile since it led to the discovery of certain aspects which would not have come out if the work was done with taluk as the unit of study. The task was stupendous but the rewards were adequate. Secondary data have also been collected from a number of sources. The sources are acknowledged wherever the data were used. Certain discrepancies were encountered while processing, but such data were either discarded or clarifications were sought from the concerned authorities.

Besides collecting secondary data, it was felt necessary to conduct a number of field surveys and observational tours to gather primary information. A detailed socio-economic survey of thirty-seven sample villages was undertaken. Crucial data pertaining to land holdings, agriculture, housing, caste composition etc. were obtained through questionnaire based survey. The sample has been so picked up as to be within the limits of our personnel time and also statistically correct. A survey of administrative and commercial establishments in urban areas of the command area was also conducted with the purpose of eliciting information regarding urban centre's functional characteristics.

The Study Area

The command area of Sriramsagar project falls in four districts namely Adilabad, Nizamabad, Karimnagar and Warangal. There were 11 taluks in the command area in 1971. Subsequently the number of taluks has gone up to 16 by 1981 as a result of a reorganisation. In 1985, the mandals were established in their place. A detailed list of the taluks and mandals is given in Table 1.1.

PHYSICAL SETTING

Relief

There are three ranges of hills in the Sriramsagar Command Area. The first one extends from Gurrapalle in Metpalli taluk in a north-easternly direction as far as Jagtial to Vemalkurthi near Godavari river. Another range known as Shanigaram range runs parallel to the first range from Shanigaram and Mallangur located in bordering Karimnagar taluk. The third range originates from the Maner river valley and extends in north-easternly direction passing beyond Ramgir.

Geology*

Geologically, Sriramsagar Command Area is composed of two types of geological formations. The western part consists of peninsular granite complex with minor bands of Dharwarian rocks. The eastern part comprises of purana sedimentaries, Pakhals and Sullavmais, occurring as strips running in a north-west to south-east direction. The area also has Gondwanas with coal bearing formations.

A major position of the district is composed of peninsular granite complex consisting of pink and grey granites, the colour of which depended upon the colour of felspars contained in them. Granites are exposed in the form of sheets and dome-like on bouldary masses lying scattered in the flat undulating country. They are medium to course grained and are composed of quartz, felspars, muscovite and varying amounts of ferromagnesium minerals such as biotite and hornblende. The granites appear

* Based on Geological Survey of India reports and Meteorology Departments Statistics.

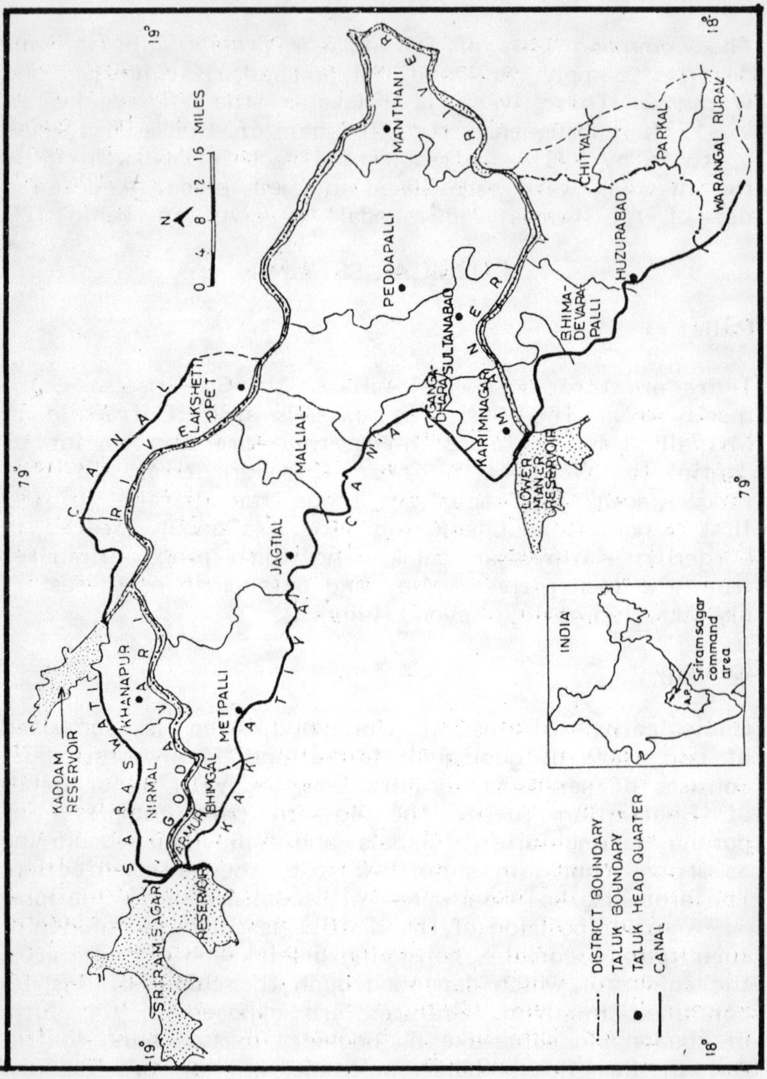

FIG. 1.1: *Sriramsagar Command Area*

1.1 TABLE 1.1 : **A Detailed List of the Taluks and Mandals**

District	Taluks (1981)	Taluks (1971)	Mandal (1985)
1	2	3	4
Adilabad	Khanapur	Lakshetipet	Luxettipet
	Lakshettipet	Khanapur	Dandepalli
			Jannaram
			Kadam
			Khanapur
			Mamda
			Laxmanchanda
			Nirmal
			Dilwarpur
Nizamabad	Armur	Armur	Balkonda
	Bhimgal		Mortha
Warangal	Chityal	Parkal	Chityal
	Parkal	Warangal	Parkal
	Warangal		Mogulapalli
	Rural		Hasanparthi
			Atmakur
			Shyamkur
			Duggondi
			Hanamkonda
Karimnagar	Metpalli	Karimnagar	Ramagundam
	Jagtial	Metapalli	Kamanpur
	Mallial	Jagtial	Manthani
	Sultanabad	Peddapalli	Kataram
	Peddapalli		Mahadevpur
	Manthani	Manthani	Mutharam
	Gangadhara	Karimnagar	Tadicherla
	Karimnagar		Mutharam
	Huzurabad		Srirampur
	Bhimdevapalli		Odela
			Jammikunta
			Kamalapur
			Elkaturthi
			Bheemadevara-palli
			Metpalli
			Ibrahimpatnam
			Mallapur

Contd.

1	2	3	4
			Raikal
			Dharmapuri
			Velgatur
			Dharmavaram
			Peddapalli
			Sultanabad
			Veenavanka
			Huzurabad
			Keshavapatnam
			Manakondur
			Maidpalli
			Koratla
			Jagtial
			Gollapalli
			Pegadapalli
			Julapalli
			Choppandi

banded or streaky before the granites were fully consolidated. They were changed to gneissic granites where they were subjected to great pressure of metamorphic agencies. Biolite granites cover wide areas and contain at places. They are typical dome shaped hills. Some of the massive outcrops of dark coloured rocks in the Sriramsagar Command Area probably represent the charnockites showing variation in the composition from acid to basic and invariably containing hypersthene with or without augite. The granite are dissected by numerous dolerite dykes ranging in thickness from a few metres to over hundred metres.

The puranas occur in a narrow strip, running roughly north-west-south-east in the eastern and central part of command area in Manthani and Peddapalli taluks. The Pakhals in this area are composed mainly of dolomites, glauconitic sandstones, limestones and shales.

The Gondwanas are fresh water sedimentary rocks deposited in steam beds on lakes and contain worable coal seams in them. They occupy a long belt running north-west to south-east along the eastern parts of the command area covering Manthani and parts of Peddapalli taluks.

Climate

The climate of Sriramsagar Command Area is characteri-

TABLE 1.2 : **Sriramsagar Command Area : Average Monthly Rainfall - 1901-80 (in mm)**

	January	February	March	April	May	June	July		Sept.	Oct.	Nov.	Dec.	Total
Karimnagar	6.1	22.1	12.9	21.1	21.8	149.6	220.2	169.9	173.2	54.9	20.6	6.6	879.0
Huzurabad	5.6	20.3	11.9	22.3	31.0	115.3	219.7	175.3	190.5	68.8	24.4	1.8	886.9
Jagtial	3.6	24.4	11.9	222.9	15.5	145.5	264.9	185.7	174.2	69.3	24.7	2.0	944.0
Peddapalli	4.6	20.6	10.2	25.9	20.6	140.2	301.7	184.1	157.2	65.5	12.7	3.3	946.6

sed by hot, summer and moderate winters. Generally there is not much of rainfall during summer and winter months. Rainfall mostly occurs during the period of south-west monsoons *i.e.* June to September.

The average annual rainfall the command area receives is moderate. However high variations can be observed from year to year, but also during the monsoon period. About 80 per cent of the total rainfall is received during the monsoon period from south-west monsoons.

The only meteorological station located in Sriramsagar command area is Ramagundam. The records of this station may be considered as fairly representative of the meteorological conditions of command area. May is the hottest month with the mean daily maximum at 42.7°C and mean daily minimum at 29.4°C. On many days is the summer, the day temperature exceeds 40°C. Day temperatures come down after the advent of south-west monsoons in June-July. By about the first week of October, the monsoon withdraws and there is a slight increase in the day temperature. However, night temperatures steadily decrease. December is the coldest month with the mean daily maximum at 29.9°C and mean daily minimum at 14.6°C.

TABLE 1.3 : **Sriramsagar Command Area : Frequency of Annual Rainfall (1901-1980)**

Range in mm	No. of years
301-400	1
401-500	2
501-600	2
601-700	4
701-800	7
801-900	24
901-1000	22
1001-1100	8
1101-1200	2
1201-1300	3
1301-1400	1
	76*

* Data not available for four years.

The night temparatures sometimes drop down to $9^{o}C$. After February temperature begins to increase.

Rivers

The Godavari river is the largest river of peninsular India with its origin at Triambak near Nasik in the Western Ghats region. The river enters Andhra Pradesh near Basar in Adilabad district. The river enters Karimnagar district near Kandukurthi in Metpalli taluk. It acts as the border between Adilabad and Karimnagar districts. It flows for about 283 kilometres in the district of Karimnagar and separates Chanda and Bastar districts of Maharashtra and Madhya Pradesh States respectively.

Another river that flows through Sriramsagar command area is River Maner. It is a tributary of River Godavari. The river originates near the village of Kalkur and traverses the district from west to east as far as Karlakunta in Manthani taluk. From there, it takes a northly course and joins the river near Mahadevpur located in the same taluk. Its length in the Sriramsagar command area is 232 kilometres and is an important source of irrigation before it formed part of Sriramsagar irrigation system.

A number of sub-tributaries popularly known as Peddavagu, Chinnavagu also traverse Sriramsagar command area. They are not perennial streams.

Mineral Resources

The important minerals that are found in Sriramsagar command area are coal, iron ore, limestone and soapstone. Coal is found in Barakar sandstones of the Gondwana formations in the eastern part of Sriramsagar command area. There are four workable seams in this area, having thicknesses ranging from 3.33 m to 6.36 m. The proved resources are approximately of the order of 150 m tonnes to a depth of 300 metres and 250 million tonnes to a depth of 450 metres. Coal is being mined in this area by Singareni Collieries Company. The Coal-fields are known as South Godavari Coal-fields or Godavari Khani coal-fields.

Many occurrences of magnetite quartz rock are located near Amberpet in Jagtial Taluk. They are thin discontinuous sheets enclosed in granite. The iron content of the magnetite quartz rock is estimated to be 35-40 per cent. The reserves are of the order of 12 million tonnes. At present there is no mining of iron ore.

More than 250 million tonnes of limestones suitable for cement manufacture are found in Sriramsagar command area. Of these one-third of the deposits occur in forested area of Peddapalli and the most of the rest in Jagtial taluk. The deposits have a high degree of calcium carbonate but a low component of magnesium carbonate.

NOTE AND REFERENCE

Rao, R.R.M., *Agrarian Development in Nagarjunasagar Command Area*, (Mimeographed paper) 1976.

TWO

Population Aspects

SRIRAMSAGAR Command Area with a population of over thiry lakhs (in 1981) is a moderately dense region of Telangana region of Andhra Pradesh State. Nearly fifteen per cent of Telangana's population and 5.6 per cent of State's population is concentrated in the command area. Sriramsagar Command Area together with Nagarjunasagar Command Area accounts for a little less than a quarter of the total population of Telangana. If the second stage of Sriramsagar and the extension of Nagarjunasagar project are completed, nearly half of the population of Telangana comes under the shadow of irrigation umbrella. The vast potential of human resources needs to be assessed and harnessed in the context of economic and social planning. Therefore, study of population characteristics is important not only in the context of overall development of the command area but also for the reason that the manpower constitutes a major input in agricultural activity. Onset of irrigation facilities and the resultant requirements need evaluation of the existing potential of human resources and estimation of future needs.

The present chapter is concerned with the analysis of major characteristics of population, namely, density, sex composition, literacy, occupational structure, dependency ratio etc. An account of urban population characteristics is also given. The characteristic features of the population of Sriramsagar Command Area are compared with those of Nagarjunasagar Command Area (left bank), as also with those of Telangana and Andhra Pradesh. Most of the aspects of population are analysed with village as the primary unit of the study, so that, variations which exist at micro-level are not omitted. Village-wise population data were collected from secondary sources and the processed data were cartographically represented. Taluk-wise figures are given in the tables accompanying the text. The major aim of the study is to identify the change in population characteristics caused by the release of waters after 1971. The collection of data pertaining to three periods of time *viz.*, 1961, 1971 and 1981, was found necessary. These three periods of time are significant, since, 1961 figures indicate the picture which existed a decade before the release of waters and 1981 picture reflects the change evident in the irrigated component of the Sriramsagar Command Area ten years after the provision of assured irrigation facilities. These two figures are to be compared with those of 1971.

Growth Pattern of Rural and Urban Components of Population

With a population over thirty lakhs in 1981, Sriramsagar Command Area accounts for 5.7 per cent of Andhra Pradesh population and 6.3 per cent of its area. This pattern conforms with that of Telangana Region (Table 2.1). However, Sriramsagar Command Area differs markedly with Nagarjunasagar Command Area, Telangana and Andhra Pradesh in regard to the growth of population. Though the population of Sriramsagar Command Area grew by 11.8 per cent during the decade 1971-81, the figure is just a little over half of the rate for the decade 1961-71. Nagarjunasagar Command Area, Telangana and Andhra Pradesh not only registered higher growth rates than that of Sriramsagar Command Area, but also improved their growth rates over the preceding decade. In the absence of detailed field investigations, it is difficult to explain the fall in the growth rate in Sriramsagar Command Area during the last censal decade. Actually, it should have been otherwise, since the provision of irrigation facilities and consequent spurt in developmental activities would have normally been expected to lead to an increase in population by in-migration.

An overwhelming proportion (84.9%) of command area's population is found in rural parts. However, there is a steady decline of the proportion of people living in rural areas over the last twenty-year period. The rural component of population shrunk to 84.98 per cent in 1981 from 93.3 per cent in 1961 (Table 2.2). The rural proportion of population declined in Telangana

TABLE 2.1 : **General Population Characteristics: Sriramsagar Command Area, Nagarjuna-sagar Command Area, and Andhra Pradesh 1961 to 1981**

		Sriramsagar Command Area	Nagarjuna-sagar Command Area	Telangana	Andhra Pradesh
Area (sq. km)		17,538	7,228	1,14,700	2,76,700
% of Andhra Pradesh		6.3	2.6	41.4	100.0
Total Population	1961	22,35,230	9,09,946	127,11,785	359,83,000
	1971	27,31,305	11,21,205	158,17,895	435,02,708
	1981	30,54,890	14,65,846	201,69,804	535,92,605
as % of A.P.		5.7	2.7	37.6	100.0
Population (Decade Variation)	1961-71	22.19	23.21	24.43	20.90
	1971-81	11.84	30.73	27.51	23.19
Population density per sq. km	1961	127	126	111	130
	1971	156	154	138	157
	1981	174	201	176	195
Rural population	1961	20,89,208	7,95,342	102,62,550	287,09,000
	1971	24,80,489	10,11,326	14,96,516	351,00,000
	1981	25,92,448	12,71,810	150,93,757	411,34,896
Urban population	1961	1,38,303	1,14,604	23,36,510	62,74,000
	1971	2,50,816	1,09,879	33,21,379	84,03,000
	1981	4,62,442	1,94,036	50,76,047	124,57,709

Source : Census of India.

TABLE 2.2 : **Sriramsagar and Nagarjunasagar Command Areas : Rural-Urban Components of Population -- 1961 to 1981**

	% Rural Component			% Urban Component			Total
	1961	1971	1981	1961	1971	1981	
Sriramsagar Command Area	93.3	90.8	84.9	6.7	9.2	15.1	100
Nagarjunasagar Command Area	87.5	90.2	86.8	12.5	9.8	13.2	100
Telangana	80.7	79.0	74.8	19.3	21.0	25.2	100
Andhra Pradesh	82.6	80.7	76.8	17.4	19.3	23.2	100

and in Andhra Pradesh also, but the magnitude of the fall is not as much as in Sriramsagar Command Area.

As against the trend of decline in rural component of population, the proportion of people inhabiting in urban areas has been steadily rising in Sriramsagar Command Area. Here again, this trend is in contrast to the not very phenomenal growth of urban component in Telangana and Andhra Pradesh. It is evident from Table 2.3 that both the command areas recorded phenomenal

TABLE 2.3 : **Sriramsagar and Nagarjunasagar Command Areas, 1961-71, 1971-81 : Per cent Decade Variation of Urban and Rural Populations**

	% Decade Variation					
	Rural Population		Urban Population		Total Population	
	1961-71	1971-81	1961-71	1971-81	1961-71	1971-81
Sriramsagar Command Area	18.7	4.5	81.4	84.4	22.2	11.8
Nagarjunasagar Command Area	27.2	25.8	-4.1	76.6	18.9	30.7
Telangana	21.8	20.8	42.0	51.8	24.4	27.5
Andhra Pradesh	18.2	17.2	33.9	48.3	20.9	23.2

growth in respect of urban population, whereas Telangana and Andhra Pradesh's urban populations' growth, though significant when compared to rural population is nowhere near the figures for the command areas. It could perhaps be inferred that the fast developing segments of the state in the agricultural sector require greater infrastructure facilities which obviously, could only be located in urban areas of small, medium and large sizes.

The urban population in Sriramsagar Command Area increased steeply during the last decade (by 84.4 per cent) at the cost of increase in rural population which recorded a meagre 4.5 per cent increase. As has been observed earlier the phenomena of less than the expected increase of rural population in an irrigated farming area demands a complete investigation of the factors that are responsible for the emergence of such a trend. However, it may not be out place to mention here, that many land owners, except the small farmers, set up alternate new homes in urban areas either in the command area or outside it, due to a fear complex generated by the naxal movement. This aspect will be elaborated in the forthcoming chapters.

Population Density

Many previous investigations have indicated that geographical, economic and social factors have largely been responsible for the existing patterns of spatial distribution of population in an area. It would be worthwhile to discuss the distinct spatial variations in the distribution and density of population brought about by geo-economic factors. In the past, the population with subsistence agriculture as its mainstay, was concentrated in the river belts. But with the introduction of modern irrigation methods, the distributional pattern of population underwent a radical change in many areas of the world. The newly emerged patterns have been consolidated by the construction of transportational facilities in recent times. It is also established that productivity of the land and the security in agriculture have great influence upon the distribution and density of population. The population density can be calculated in two ways--physiological and rural. Physiological density indicates the ratio between rural population and arable land; whereas, the rural density is computed by working out the quotient between rural population and total land area. Both the methods have their advantages as well as disadvantages.

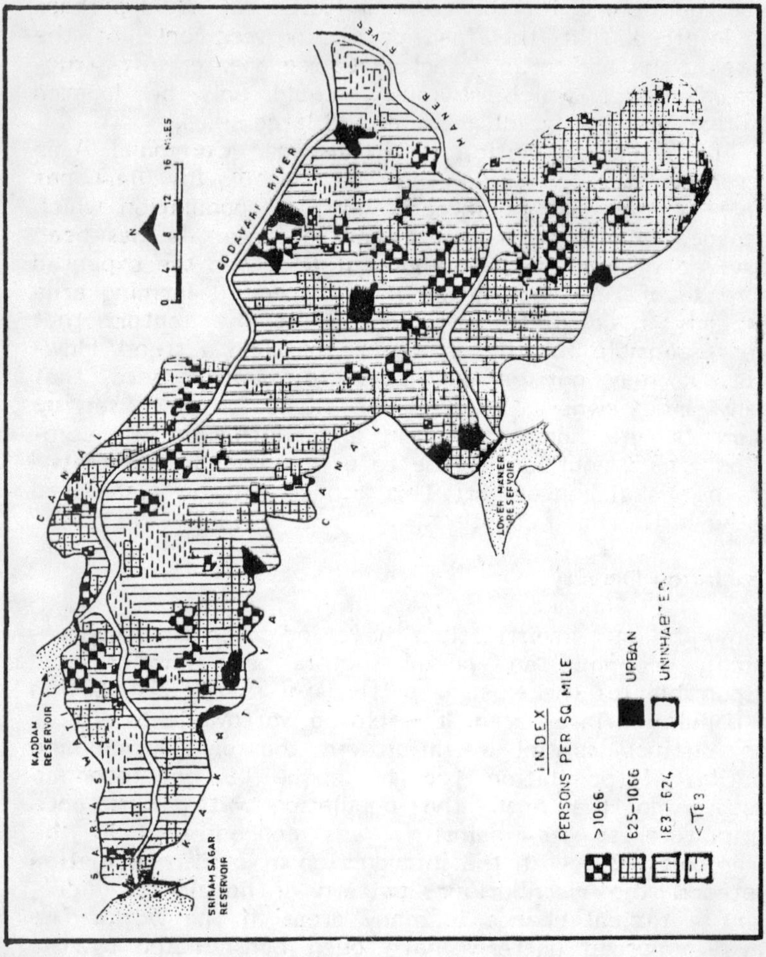

FIG. 2.1:
Sriramsagar Command Area: Population Density 1981

However, it has generally been observed that density of population is closely related with the food production possibilities, where, irrigation facilities are being developed and are likely to grow in future. It is, therefore, vital to examine the hypothesis that increase in prosperity of an area leads to relatively higher density of people.

In the case of Sriramsagar Command Area, the density and distribution pattern of population is more or less in conformity with those of Nagarjunasagar Command Area and the rest of the State. There is no valid premises to observe that density pattern in Sriramsagar Command Area, as a whole, stands out markedly in contrast to other areas of the State with similar economic types and similar economic structuring. In the distant past, the passage of Godavari river through the Command Area had not made any significant impact on the pattern of population distribution. This was because of the factor that there were no means of making use of river waters for agricultural purpose, on a large scale. Since the river had its middle course only in the area, the region on the two sides of the river was impossibly high for any conventional type of lift irrigation. Only after the advent of the new dam building technology, it has now become possible to harness the river.

Generally speaking, the areas around the urban centres and the area in the north of Sriramsagar Command Area are densely populated when compared to other areas. Only 19.5 per cent of the total area has density of more than thousand persons per square km. About half of the command area has density of less than 185 persons per sq. kilometre. This indicates that population is not evenly distributed. Although the phenomenon of uneven distribution of population can be noticed, a closer analysis of village-wise density map reveals certain trends and patterns with regard to variation in density over a period of time. A cursory examination of Table 2.4 indicates that the areas which have been benefited by irrigation facility for a longer period of time, have experienced higher percentage decade variation of density than the areas which have, either, not received water so far, or have received only recently. For example, Jagtial and Sultanabad taluks which were provided irrigation facilities around the year 1971, registered high percentage decade variations of 24.7 and 30.7 during 1971-81 decade, as against relatively low percentage decade variation

TABLE 2.4 : **Sriramsagar Command Area : Density of Population and Decade Variation (Taluk-wise) -- 1961 to 1981**

Sl. No.	Taluk	Persons per sq. mile			% Decade Variation	
		1961	1971	1981	1961-71	1971-81
1.	Metpalli	390	535	655	37.7	22.3
2.	Jagtial	353	419	522	18.4	24.7
3.	Mallial	353	419	522	18.4	24.7
4.	Sultanabad	394	503	658	27.7	30.7
5.	Peddapalli	394	503	658	27.7	30.7
6.	Manthani	118	142	172	21.4	19.3
7.	Gangadhara	466	542	696	16.4	28.2
8.	Karimnagar	466	542	696	16.4	28.2
9.	Bhimadevara-palli	497	563	658	17.5	16.5
10.	Huzurabad	497	563	658	17.5	16.5
11.	Chityal	328	406	494	23.8	21.1
12.	Parkal	328	406	494	23.8	21.1
13.	Warangal (Rural)	480	566	N.A.	17.8	N.A.
14.	Nirmal	243	304	N.A.	25.5	N.A.
15.	Khanapur	178	211	262	20.0	24.2
16.	Laxettipet	168	238	347	41.4	46.5

of 18.4 and 27.7 respectively during the decade prior to it. The taluks which are yet to receive water or which have been provided with irrigation facility only recently are characterised by either a declining trend of per cent decade variations of density from 1961-71 to 1971-81 or a stagnant state. Bhimadevarapalli, Huzurabad and Chityal taluks which come under the unirrigated sector of Sriramsagar Command Area reflect this aspect.

Sex Composition

The proportion of males and females in a population is normally referred to as sex ratio. There are many methods of computing sex ratio. The most widely used method is the calculation of number of females per thousand males. Variation in sex ratios is often indicative of certain aspects of social and economic setting of the areas. Low sex ratios *i.e.* less number of females than males results mainly from predominance of male births high rate of female mortality and male selective in-migration. Low sex ratio is found in areas experiencing male influx such as those near the major urban centres, on the areas which had some construction project under progress.

Provision of assured irrigation facilities often leads to changes in the cropping system from dry crop cultivation to intensive wet crop cultivation. Normally, the wet crop that is suitable for the area is rice. The increased cultivation of rice in the command area demands availability of female labour which is required in many important operations of rice cultivation specially in levelling, transplanting, weeding and harvesting. These activities are done by the female labour only as a matter of tradition and also economic convenience. It can, therefore, be assumed that the sex ratio in Command Areas of irrigation projects will be relatively in favour of females than in the irrigated dry crop areas. However, it does not mean that irrigation is the sole determinant of sex-ratio. It is just one of the many factors which cause the imbalance in sex composition. There may be a number of other factors which have the same or greater influence in determining sex ratios. An examination of Table 2.5 substantiates this inference. Sriramsagar Command Area has more females per thousand male population than Telangana and Andhra Pradesh. The low female proportion in Telangana as a whole is under-

TABLE 2.5 : **Sriramsagar and Nagarjunasagar Command Areas : Sex Ratio and Literacy Ratio—1961 to 1981**

	Sex Ratio Females per 100 males			Literacy Ratio			Percentage decade Variation of literate population	
	1961	1971	1981	1961	1971	1981	1961-71	1971-81
Sriramsagar Command Area	982	973	984	12.7	15.0	21.2	44.7	57.7
Nagarjunasagar Command Area	943	953	955	15.8	18.7	25.9	46.1	80.7
Telangana	975	968	971	17.3	20.7	26.2	48.6	61.5
Andhra Pradesh	981	977	976	21.2	24.6	29.7	40.1	49.0

Source : Census of India.

standable since, the region is mostly a dry crop growing area. The low proportion of females in Nagarjunasagar Command Area is however, contrary to the inference drawn above. In the absence of detailed investigations, it is difficult to pinpoint the underlying causes for the emergence of the pattern. However, marked rise in proportion of female population can be observed in Nagarjunasagar Command Area also from 1961 to 1981.

The distributional pattern of sex ratio as can be seen from Table 2.6 and Fig. 2.2 substantiate the observation made earlier. The irrigated component of the command area specially the taluks of Jagtial, Mallial, Sultanabad, Khanapur have relatively higher proportion of female population than the unirrigated component of the command area.

To sum up, it may be said that relatively higher proportion of females are found in rice growing command areas of irrigated projects. Two possible factors can be investigated. Does this pattern emerge because of in-migration of female labour from other areas? However, as a matter of tradition, female labour does not migrate without being accompanied by their male counterparts. Secondly it is possible that since, in rice cultivation more of female labour is required than male labour, the men migrate to work places outside the command

TABLE 2.6 : **Sriramsagar Command Area : Sex Ratio, Literacy Ratio and Growth of Literate Population (Taluk-wise for 1961, 1971 and 1981)**

Sl. No.	Taluk	Sex Ratio Females per 100 males			Literacy Ratio			Percentage decade variation of literate population	
		1961	1971	1981	1961	1971	1981	1961-71	1971-81
1.	Metpalli	1064	1005	1069	11.3	12.2	17.8	49.8	78.1
2.	Jagtial	1029	1008	1009	11.3	13.8	20.9	50.0	67.7
3.	Mallial			1043			14.5		
4.	Sultanabad			978			16.4		
5.	Peddapalli	989	983	982	11.5	14.4	25.0	59.8	101.6
6.	Manthani	964	956	974	9.8	10.9	17.5	34.4	71.5
7.	Gangadhara			101			20.1		
8.	Karimnagar	974	975	950	16.6	20.7	38.1	45.5	76.9
9.	Bhimadevar-palli			99.1			19.1		
10.	Huzurabad	966	969	962	12.6	14.41	25.0	34.3	82.8
11.	Chityal			977			13.0		
12.	Parkal	954	954	973	11.2	11.1	18.9	22.2	84.7
13.	Warangal Rural	960	947	991	13.3	16.3	20.4	44.0	N.A.
14.	Nirmal	1031	1005	943	12.4	14.9	9.8	51.2	63.8
15.	Khanapur	931	973	986	10.4	11.5	15.7	31.0	69.7
16.	Lakshettipet	948	943	947	12.4	15.6	21.1	77.8	97.8

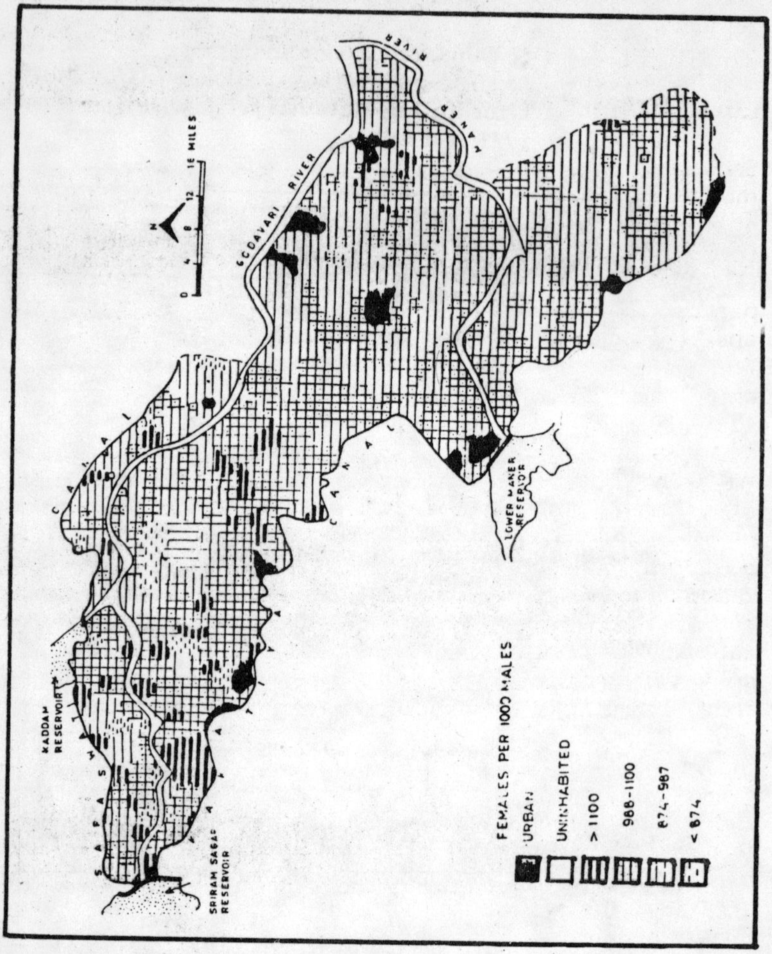

FIG. 2.2:
Sriramsagar Command Area: Sex Ratio 1981

area, thus leading to relatively higher proportion of females. How this process takes place is a debatable point in the absence of concrete evidence.

Literacy Ratio

Sriramsagar Command Area was perhaps one of the most economically backward areas of Telangana region and Andhra Pradesh State. This factor could be attributed to a very low level literacy of the people of Sriramsagar Command Area (Table 2.5). The literacy ratio was 21.2 per cent in 1981 as against 25.9 per cent, 26.2 per cent and 29.7 per cent for Nagarjunasagar Command Area, Telangana and Andhra Pradesh respectively. It is not only that the literate population during the decades 1961-71 and 1971-81 is also relatively less than the corresponding figures for Nagarjunasagar and Telangana. We can interpret the above to mean that the fruits of economic prosperity brought about by the provision of irrigation facilities have not yet ripened and led to subsidiary benefits like higher literacy. Another possible reason for lack of substantial increase in literacy level could be due to increase in the quantum of employment caused by intensive cultivation of the land. The school going children would have perhaps found work in the fields more compelling than going to school. However, it is difficult to substantiate this point without further indepth studies.

An examination of Fig. 2.4 and Table 2.6 supports the contention made above. The irrigated areas have a relatively low levels of literacy than the unirrigated parts of the Sriramsagar Command Area. The urban areas and the surrounding regions exhibit relatively higher levels of literacy. This could be due to provision of education facilities in urban areas. This aspect leads us to one very vital question. Does the provision of facilities for getting educated lead to higher level literacy? Does economic prosperity has no bearing in increasing the educational consciousness of the people? The situation as exists in Sriramsagar Command Area supports the above contentions at least superficially.

Dependency Ratio

Population in an area can be categorised into two distinct

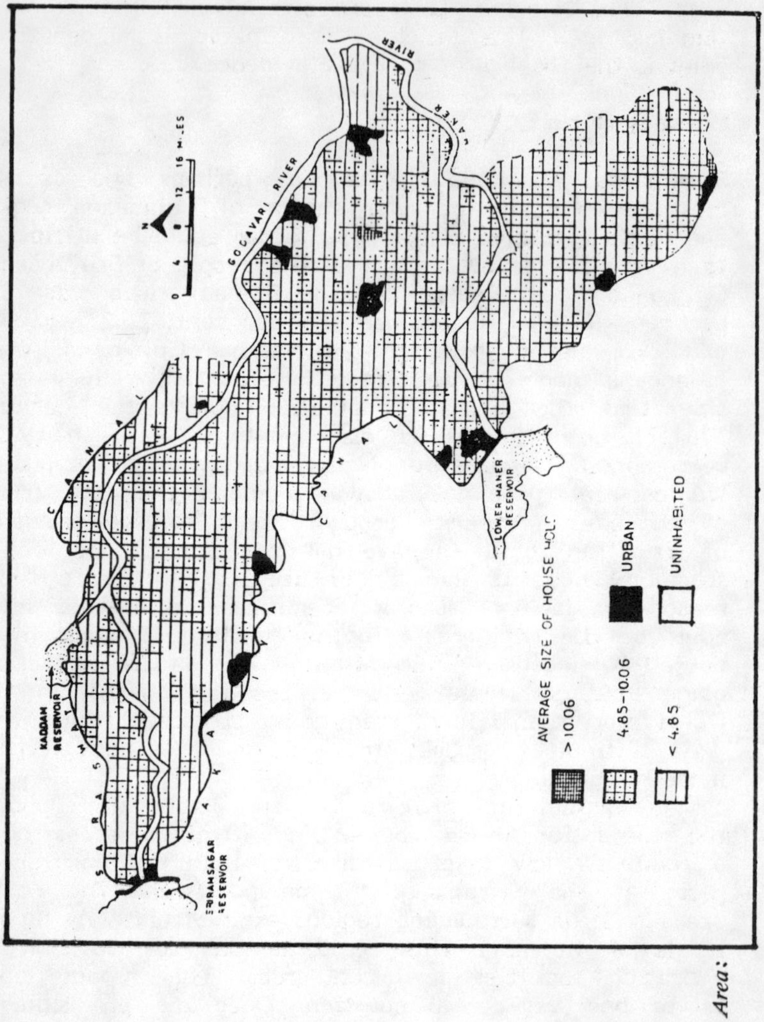

FIG. 2.3:
Sriramsagar Command Area:
Household Size 1981

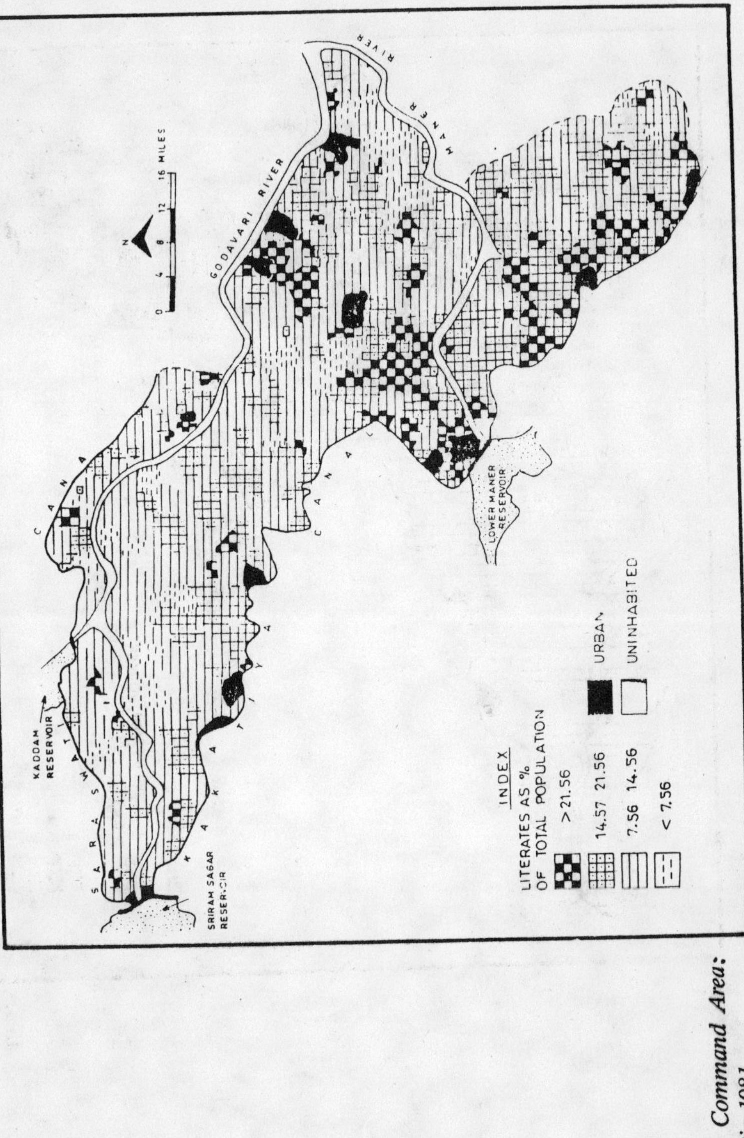

FIG. 2.4:
Sriramsagar Command Area: Literacy Ratio 1981

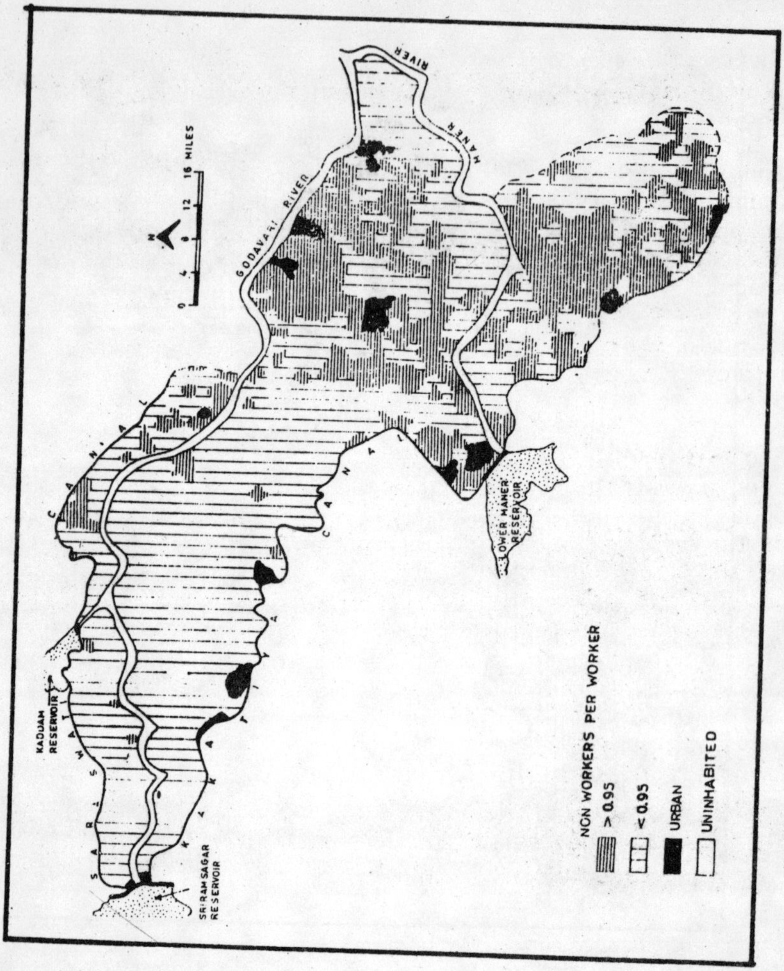

FIG. 2.5:
Sriramsagar Command Area: Dependency Ratio 1981

groups *viz.*, workers and non-workers who either, do not have jobs to do, or are too old or too young to pursue an occupation. In simple words they are economically inactive. It also includes that section of population which is still in the process of acquiring necessary skills and knowledge required for pursuing an occupation. The intricate relationship between the proportion of workers and non-workers, if expressed in the form of a ratio (number of non-workers per worker), can be termed as dependency ratio. High dependency ratio can be attributed to a number of socio-economic and cultural-demographic factors in their complex combinations. Here, the Dependency Ratio has been calculated as the number of non-workers per worker. Dependency Ratio, if properly interpreted keeping in mind the limitations imposed by certain lacunae in its computation, can be considered to reflect some important aspects of the population under study. Generally speaking, a lower value of dependency ratio is indicative of a higher degree of work opportunities and high degree of economic activeness of the population. However, this generalisation is subject to the assumption that proportion of people who are too old or too young to work, is uniform throughout the area of study. The proportion of people pursuing education also causes disequilibrium in the dependency ratio. Barring the above exceptions, it can, perhaps be considered that lowering of dependency ratio over a period of time in a given area is reflective of economic prosperity caused by provision of more job opportunities.

Sriramsagar Command Area which experienced a decrease, though insignificant, of its active population in the decade prior to the release of waters, registered substantial increase in the working population during the period 1971-81. The proportion of non-workers came down in ample measure during the decade following the release of waters, thus reversing the trend for the period 1961-71 (Table 2.7). In this respect, the situation is similar to that of Nagarjunasagar Command Area. There were only 110 dependents for every hundred workers (in 1981) in Sriramsagar Command Area, whereas the corresponding figures were 127, 130 and 137 for Nagarjunasagar Command Area, Telangana and Andhra Pradesh respectively. The relatively lower value of dependency ratio in Sriramsagar Command Area is indicative of the economic buoyancy and the magnitude of available

TABLE 2.7 : **Dependency Ratio - 1961 to 1981 : Sriramsagar Command Area, Nagarjunasagar Command Area, Telangana, Andhra Pradesh**

		Population in thousands	Workers in thousands	Non-workers in thousands	Dependency ratio (%)
Sriramsagar	1981	3054	1452	1602	1.10
Command	1971	2731	1242	1489	1.19
Area	1961	2235	1266	969	0.76
Nagarjuna-	1981	1465	645	820	1.27
sagar Comm-	1971	1121	471	650	1.38
and Area	1961	909	482	427	0.88
Telangana	1981	20168	8769	11400	1.30
	1971	15817	6771	9046	1.33
	1961	12711	6818	5893	0.86
Andhra	1981	53592	11606	10986	1.37
Pradesh	1971	43502	18005	25497	1.41
	1961	35983	18663	17320	0.92

job opportunities.

If we examine the pattern of intra-command area dependency ratio, it is quite evident that irrigated part of command area on the whole, has less than one dependent for each worker. Sixty-five per cent of the command area, most of which is irrigated, has a dependency ratio of less than 0.95. The areas surrounding towns, both small and medium, have a higher value of dependency ratio. The low magnitude of dependency ratio in the irrigated component can be attributed to the booming agricultural activity which, in turn, throws open a number of job opportunities, mostly in agricultural sector itself, but sometimes in non-agricultural sector as well. As in the case of other agriculturally important parts of the country, intensive irrigated agriculture requires more labour. It is common to observe all members of the family working either in the field owned by them or in the fields of others for a wage. Higher dependency ratio in urban areas and in their surroundings could be

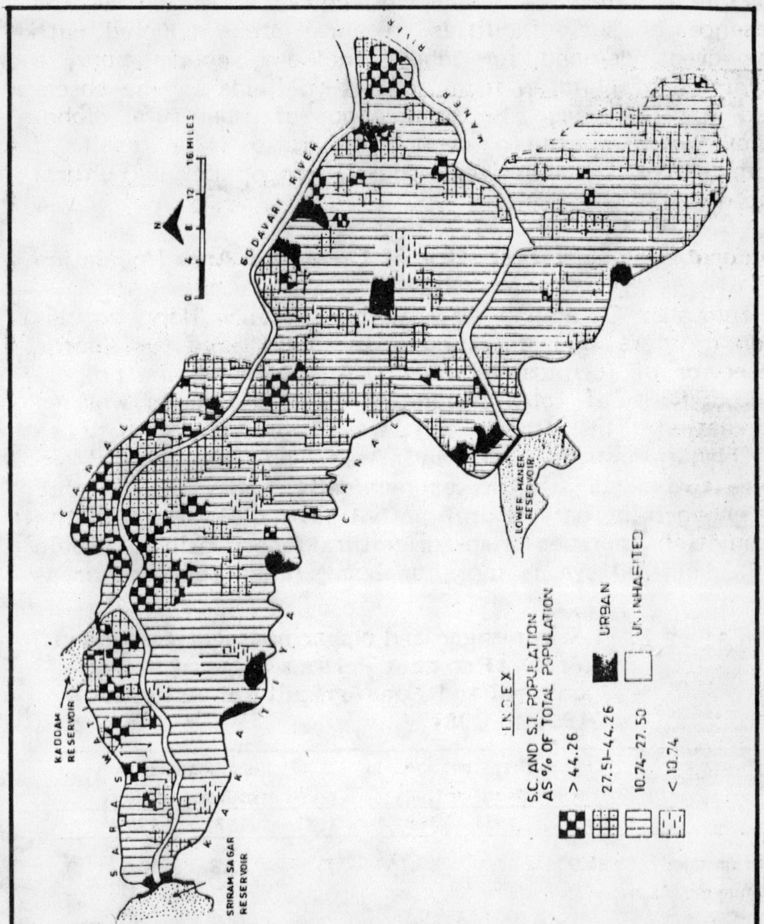

FIG. 2.6:
Sriramsagar Command Area:
S.C. & S.T. Population 1981

due to the availability of educational facilities in them, to which children are sent to acquire education. In the absence of such facilities in rural areas coupled with the great demand for labour including child labour, a diversion of children from schools to fields can be observed in rural areas. The phenomenon of urban-rural dichotomy with regard to dependency ratio is a result of complex relations between social, economic and cultural factors.

Occupational Characteristics of Command Area Population

Sriramsagar Command Area with a total working population of more than fourteen lakhs (1981) is a work force reservior of formidable volume. An overwhelming proportion (93.3%) of total population is classified as working population. This figure compares favourably with those of Nagarjunasagar Command Area and Telangana. More than two-thirds of the economically active work force is engaged in agricultural activities and the rest of the population pursues non-agricultural occupations (Table 2.8). This pattern is more or less similar to other areas

TABLE 2.8 : **Sriramsagar and Nagarjunasagar Command Areas : Per cent Persons engaged in Agricultural and Non-Agricultural Activities - 1961 to 1981**

	% Workers engaged in agricultural activity			% Workers engaged in non-agricultural activity		
	1961	1971	1981	1961	1971	1981
Sriramsagar Command Area	64.9	67.2	69.7	35.1	32.8	30.3
Nagarjunasagar Command Area	64.4	74.0	75.9	32.6	26.0	24.1
Telangana	68.0	67.5	67.5	33.0	32.5	32.5
Andhra Pradesh	70.6	70.1	69.3	29.4	29.9	30.7

in the state. However, there is a marked increase in the proportion of workers engaged in agricultural activities in both the command areas, over the twenty years period *i.e.* 1961-81. Consequently, the command areas registered decline in the proportion of workers engaged in non-agricultural activities during the same period. This trend in the command areas in contrast to that existing in the state which experienced either decrease or stagnancy of per cent workers in agricultural activity and *vice versa* in non-agricultural activities. The substantial increase of workers in agricultural activities in the command areas can be attributed to the increased scope in and profitability of agriculture, caused by the release of waters. It can also be said that this pattern could also be due to a change in cropping system from extensive dry farming to a labour intensive irrigated farming.

An examination of occupational structure by listed categories (Tables 2.9 to 2.11) reveals that the command areas, considered as a whole, did not register increase in the 'cultivators' category. This is all the more noteworthy since the benefits of the provision of irrigation facility and development in the agricultural sector used to be considered and still supposed to be considered as an instrument of ushering in an era of economic quality. It clearly indicates that policy of encouraging owner cultivators has not really succeeded as it was planned to be. It is reflective of the failure of the so-called progressive tenancy laws and land reforms. The 'why' and 'how' factors of this trend will be discussed in the following chapters. Another point which emerges out of a coursory look at occupational categories proportions is that the share of primary activities is still very high. Increased agricultural activity in the command areas as a whole has not led to the emergence of any significant improvement of the secondary sector. The additional agricultural production and the resultant monetary benefits did not lead to any appreciable investment in enlargement of industries, despite the availability of raw materials for setting up agro-based industries in the command areas. It can only be inferred that available surplus income is not being channellised in the area, but siphoned off to other areas, probably to the large cities. However, some development in secondary activities is discernible in the irrigated component of the command area specially durinq the last few years.

TABLE 2.9 : Sriramsagar and Nagarjunasagar Command Areas : Occupational Structure As % Total Workers

	Cultivators		Agricultural Labourers		Livestock		Household industry		Other than household		Construction		Trade and Commerce		Transport storage and communication		Other services	
	1961	1971	1961	1971	1961	1971	1961	1971	1961	1971	1961	1971	1961	1971	1961	1971	1961	1971
Sriramsagar Command Area	40.4	34.5	24.4	34.7	3.0	5.6	13.9	7.5	2.8	4.5	1.0	2.3	3.0	3.1	0.4	1.2	11.0	6.5
Nagarjunasagar Command Area	36.7	32.4	30.7	41.5	2.0	4.5	4.1	4.0	1.7	2.1	3.0	3.7	3.4	4.0	0.6	1.0	8.5	6.5
Telangana	41.8	35.0	25.2	32.5	3.0	4.8	10.4	5.2	3.7	4.6	1.5	1.9	4.7	5.2	1.1	2.1	9.8	8.1
Andhra Pradesh	42.0	32.2	28.6	37.9	3.0	3.8	9.7	1.9	2.6	4.2	1.2	1.6	4.3	5.5	1.3	2.2	9.3	7.8

TABLE 2.10 : **Sriramsagar Command Area: Occupational Structure (Taluk-wise), 1971**

Sl. No.	Taluk	% Workers Engaged in the Activity									
		Cultivators	Agricultural Labourers	Livestock forestry, mining, quarrying, etc.	Household industry	Other than household	Construction	Trade and Commerce	Transport storage and Communication	Other services	Total
1.	Metpalli	38.9	19.2	2.6	9.6	9.6	11.4	2.8	0.6	4.7	100
2.	Jagtial	37.1	32.9	3.8	8.4	4.8	2.5	3.2	0.9	6.1	100
3.	Peddapalli	29.0	35.5	6.0	7.3	3.9	1.1	3.3	1.6	7.8	100
4.	Manthani	31.0	46.3	6.7	3.7	2.8	1.0	2.2	0.6	5.6	100
5.	Karimnagar	34.8	34.0	3.6	8.9	3.6	1.3	4.0	1.4	8.2	100
6.	Huzurabad	35.4	37.8	4.1	9.5	2.6	1.1	2.7	0.8	5.9	100
7.	Parkal	32.2	44.3	4.4	7.2	3.9	0.6	1.8	0.2	5.7	100
8.	Warangal Rural	34.5	38.7	3.8	6.5	3.8	1.3	2.6	2.5	5.7	100
9.	Nirmal	26.8	33.2	5.1	8.0	11.2	3.1	5.1	1.4	9.3	100
10.	Khanapur	39.0	39.0	3.3	3.8	3.4	1.8	3.0	0.3	5.9	100
11.	Laxettipet	30.7	19.1	2.6	3.0	2.1	0.6	2.4	1.1	4.6	100
	Command Area	34.5	34.7	5.6	7.5	4.5	2.3	3.1	1.2	6.5	100

TABLE 2.11 : **Sriramsagar Command Area : Occupation Structure - Taluk-wise - 1981**

Sl. No. Taluk	% Main Workers Engaged in the Activity				
	Cultivators	Agricultural labourers	Household industry	Other workers	Total Main workers
1. Metpalli	39.6	23.8	16.6	20.0	100
2. Jagtial	39.6	31.0	9.4	20.0	100
3. Mallial	41.8	33.8	9.0	15.4	100
4. Sultanabad	35.0	45.8	5.7	13.6	100
5. Peddapalli	23.1	34.0	4.8	38.1	100
6. Manthani	28.4	50.0	3.6	18.0	100
7. Gangadhara	35.4	35.6	1.5	16.5	100
8. Karimnagar	4.9	27.3	7.0	40.8	100
9. Bhimdevarpalli	37.8	40.9	8.4	12.9	100
10. Huzurabad	30.8	41.6	9.0	18.6	100
11. Chityal	39.6	42.5	5.5	12.4	100
12. Parkal	32.3	42.2	8.4	17.1	100
13. Warangal Rural	31.2	42.8	8.6	17.3	100
14. Nirmal	38.7	25.6	16.8	18.9	100
15. Khanapur	40.6	4.5	5.9	13.0	100
16. Lakshettipet	24.2	35.0	4.9	35.9	100
Command Area	32.3	38.1	10.5	19.1	100

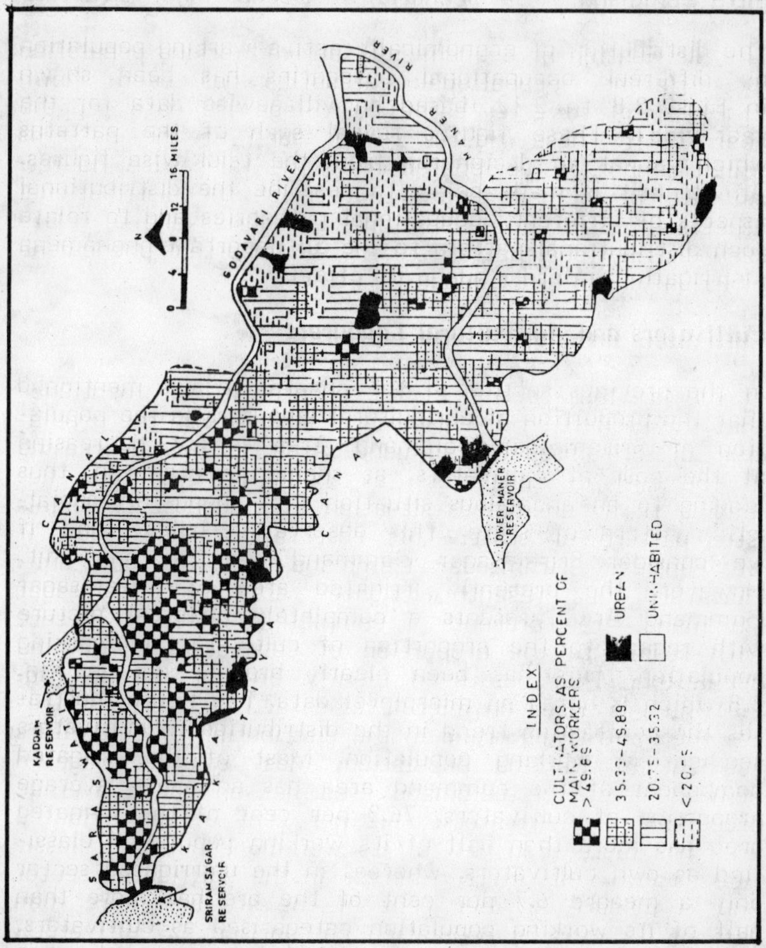

FIG. 2.7: *Sriramsagar Command Area: Cultivators 1981*

Intra-Command Area Scenario of Occupational Structure

The distribution of economically active working population by different occupational categories has been shown in Figs. 2.8 to 2.12, based on villagewise data for the year 1981. These figures reveal such of the patterns which cannot be deciphered from the taluk-wise figures. An attempt is made below to describe the distributional aspects of different occupational categories and to relate such distributional pattern to the all important phenomena of irrigation, in order guage its effect.

Cultivators and Agricultural Labourers

In the previous section of the chapter it was mentioned that the proportion of agricultural labourers in the population of Sriramsagar Command Area is not decreasing at the cost of cultivators, as should be expected, thus leading to an anomalous situation in a supposedly socialistic pattern of state. This observation holds good if we consider Sriramsagar Command Area as one unit. However, the presently irrigated area of Sriramsagar Command Area presents a completely different picture with regard to the proportion of cultivators in working population. This has been clearly brought out in Fig. 2.8 which is based on microlevel data. The figure illustrates the contrasting trend in the distribution of cultivators segment of working population. Most of the irrigated component of the command area has an above average proportion of cultivators. 76.2 per cent of the irrigated area has more than half of its working population classified as own cultivators, whereas in the unirrigated sector only a meagre 6.7 per cent of the are has more than half of its working population categorised as cultivators. A village sample survey conducted by us corroborates this fact. It can, therefore, be surmised that the provision of irrigation facilities has led to a change in the agricultural land tenure system. A more and more of marginal farmers and small farmers are no more doubling as agricultural labourers, since the benefits from their small holdings have become sufficient to sustain them. It can also be mentioned that there has been an appreciable change in farming techniques leading to increase in production.

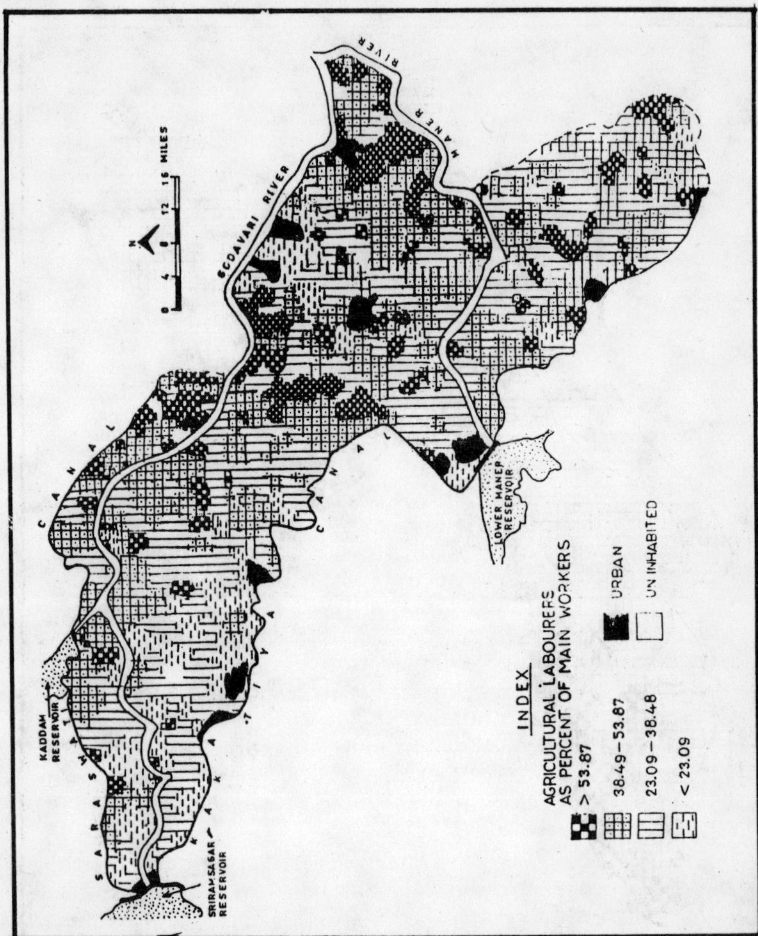

FIG. 2.8:
Sriramsagar Command Area:
Agricultural Labourers 1981

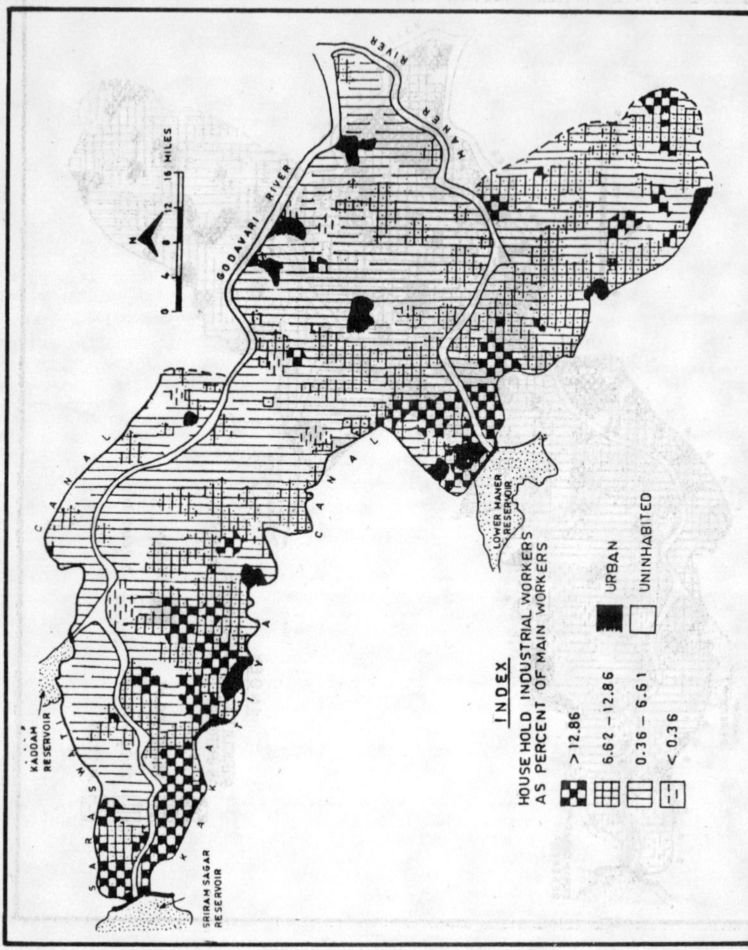

FIG. 2.9: *Sriramsagar Command Area: Workers in Household Industry 1981*

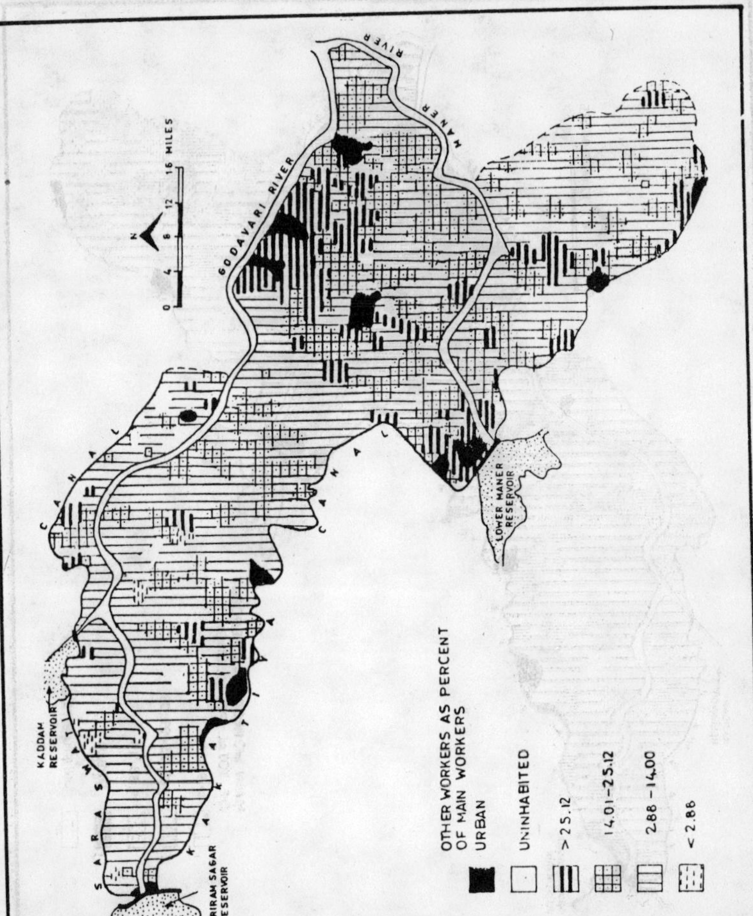

FIG. 2.10:
Sriramsagar Command Area :
Other Workers 1981

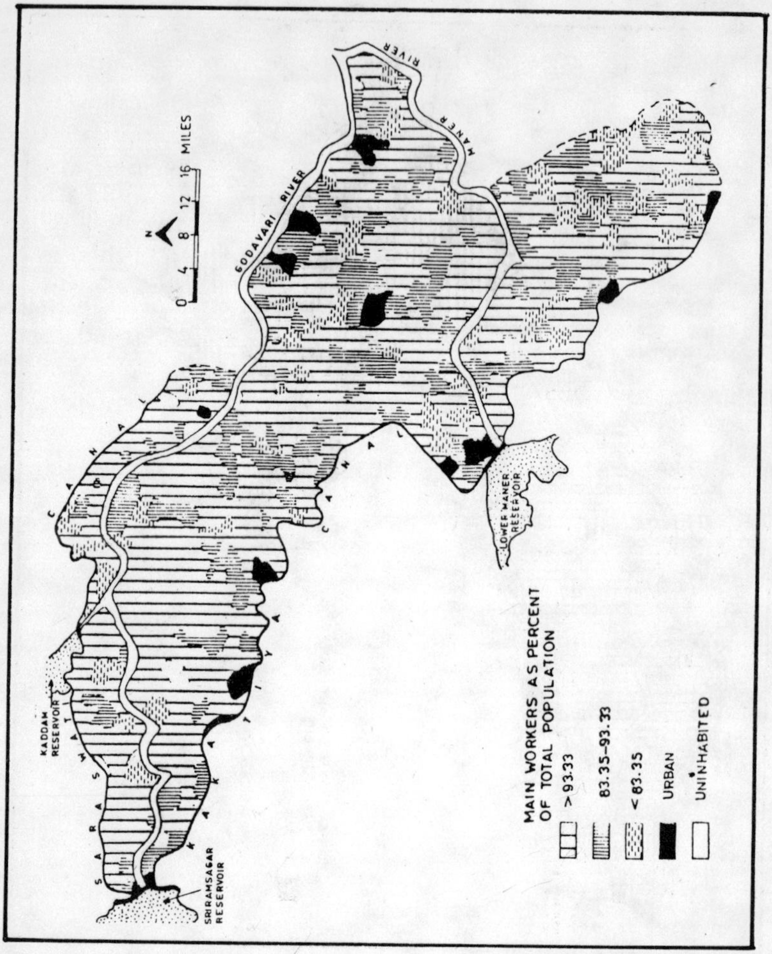

FIG. 2.11: *Sriramsagar Command Area: Main Workers 1981*

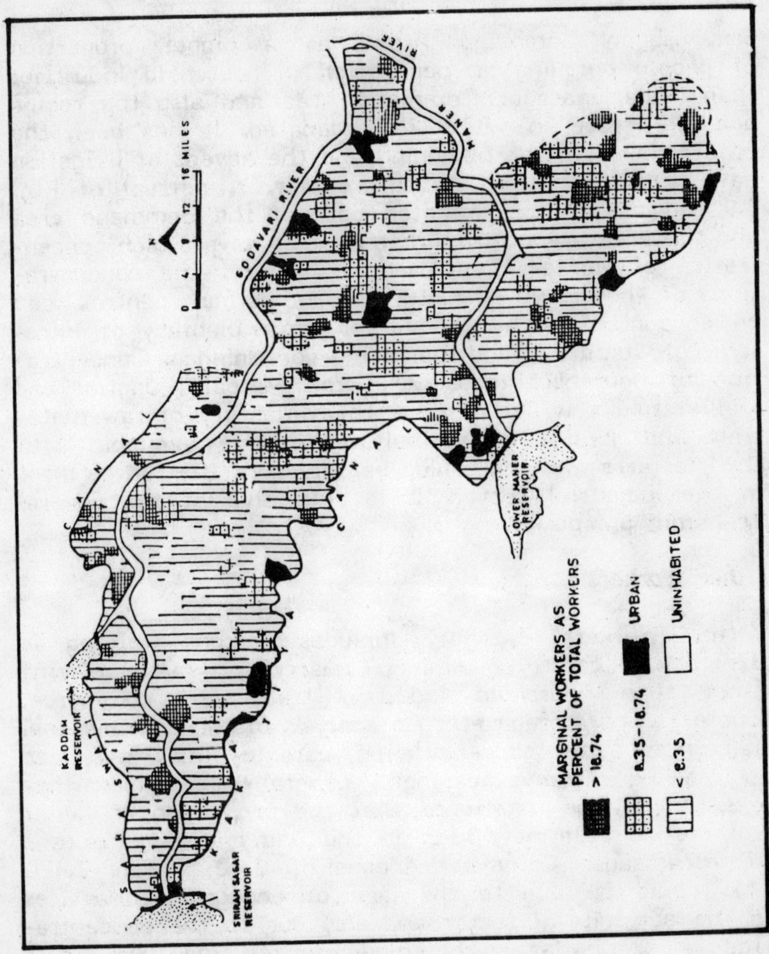

FIG. 2.12: *Sriramsagar Command Area: Marginal Workers 1981*

House Industry

Sriramsagar Command Area has a higher proportion of people pursuing an occupation in household industries than Nagarjunasagar Command Area and also the region and the state in which it is located. It has been the case right from the beginning and the advent of irrigation gave a further fillip to this sector. A persual of Fig. 2.9 shows that the irrigated part of the command area and the area around the urban centres have a high concentration of workers in household industry. The concentration of household industry around urban centres can be accounted for by the factor of availability of infrastructure facilities and marketing convenience. Concentration of household industry in the Metpalli, Jagtial and Mallial taluks is perhaps due to availability of raw materials and surplus of agricultural income available with the farmers which could be diverted for investment in the industrial sector. It is reflective of the general economic prosperity.

Other Workers

'Other Workers' category includes persons working in livestock, forestry, mining, quarrying, manufacturing other than household industry, trade and commerce, transport and storage etc. An analysis of the distributional pattern of this all embracing category may not lead to drawing of any meaningful generalisations. Nevertheless, it may be mentioned that the proportion of 'other workers' is relatively high in the unirrigated component of Sriramsagar Command Area (Fig. 2.10 to Fig. 2.12). This could be due to the lack of enough opportunities in the agricultural sector and also due to the concentration of some large and medium-sized industrial units in the agricultural sector and also due to the concentration of some large and medium sized industrial units in the north-eastern part of the command area. The urban areas and their surroundings, obviously, have high proportion of workers under this category.

Correlation between Selected Population Attributes

Pearsons product moment correlation was computed between various attributes of population in Sriramsagar

TABLE 2.12 : **Sriramsagar Command Area : Population Characteristics Correlation Matrix - 1981**

	Population Density	Sex Ratio	% Literacy	% SC.,ST. population	% Cultivators	% Agricultural labourers	% Other workers	Dependency Ratio	% Household industrial workers
Population Density		+ 0.21	+ 0.42	+ 0.11	+ 0.31	- 0.23	- 0.31	+ 0.22	+ 0.45
Sex Ratio	+ 0.21	-	- 0.31	+ 0.41	+ 0.39	+ 0.37	+ 0.12	+ 0.45	+ 0.35
% Literacy	+ 0.42	- 0.31	-	- 0.64	+ 0.22	- 0.21	+ 0.62	+ 0.41	+ 0.45
% S.C. and S.T. Population	+ 0.11	+ 0.41	- 0.64	-	+ 0.31	+ 0.63	- 0.14	+ 0.04	+ 0.10
% Cultivators	+ 0.31	+ 0.39	+ 0.22	+ 0.31	-	- 0.51	- 0.31	- 0.21	+ 0.30
% Agricultural labourers	- 0.23	+ 0.37	- 0.21	+ 0.63	- 0.51	-	+ 0.34	- 0.20	- 0.02
% Other workers	- 0.31	+ 0.12	+ 0.62	- 0.14	- 0.31	+ 0.34	-	+ 0.41	+ 0.12
Dependency Ratio	+ 0.22	+ 0.45	+ 0.41	+ 0.04	- 0.21	+ 0.20	+ 0.41	-	+ 0.23
% Household industrial workers	+ 0.45	+ 0.35	+ 0.45	+ 0.10	+ 0.30	- 0.02	+ 0.12	+ 0.23	-

Command Area. Village was considered as a basic unit. The results derived are given in the matric (Table 2.12).

Population density has significant, but not high, positive correlation with per cent workers in household industry, literacy and per cent cultivators. Sex ratio is positively correlated to some extent with per cent Scheduled Castes and Scheduled Tribes population, per cent agricultural labourers and the dependency ratio. Literacy is high in areas where other service workers form substantial proportion of population and where dependency ratio is also moderately significant. However, it is negatively correlated with per cent agricultural labourers and per cent cultivators is negatively correlated with per cent agricultural labourers and per cent other workers. Other workers is positively correlated with the dependency ratio.

THREE

Agricultural System

THE working of the agricultural system in a region is dependent on biotic and abiotic factors. Agricultural system deals with living organisms like plants and animals but depends on the natural resources like land, water and energy. However, the human element should also be considered since the farmer is the pivot of the agricultural activity. Discussion of the biotic factors in agricultural system should better be left to agricultural scientists. The social scientists, specially the geographers, and the economists are the competent persons to study and analyse the abiotic component of agricultural system. The present chapter has been organised as to cover the three main aspects of agricultural system namely, land-use, cropping pattern and agricultural efficiency.

Agriculture is the mainstay of the people inhabiting Sriramsagar Command Area as is the case in other parts of India. In the year 1985, over 69.7 per cent of work force in the command area were either owner cultivators of the land or agricultural labourers working on other peoples land. In the absence of any other major secondary and tertiary activities, agriculture is, and will continue to be, the main occupation of the people, forming the backbone of the region's economy. The release of waters and provision of assured irrigation facilities will further strengthen the agricultural component of the economy.

Land Utilization

Any worthwhile comprehensive study of a region with planning implications should surely concern itself with an analysis of the land utilization, is also a pre-requisite for rational land-use planning. An insight into the direction and distribution of different land utilization categories reveals whether there exist any regional ecological balances or imbalances. Since land utilization in a region reflects the physical, social and economic environment in which it lies, a comparison of land utilization statistics with those of other similar regions and the state gives a base for formulating a balanced land-use development plan. It also indicates whether in a particular area optimum use of the land is made or not.

Before delving deeper into an analysis of land utilization pattern of Sriramsagar Command Area, a note of caution has to be struck with regard to data availability and the deficiencies that exist in it. Land utilization data are available for the nine categories given below:

1. *Forests* : This category includes all actually forested areas or land-classed or administered as forests under any legal enactment dealing with forests, whether state-owned or private.

2. *Land Put to Non-agricultural Uses* : This stands for all lands occupied by road, buildings, industrial undertakings, grave-yards, water bodies etc., and all other land put to any other non-agricultural uses.

3. *Barren and Uncultivable Land* : This covers barren and uncultivable lands like mountains, deserts etc. Lands which cannot be brought under cultivation except at a high cost.

4. *Permanent Pastures and Grazing Lands:* This covers all grazing lands whether or 'not they are permanent pastures and meadows.

5. *Miscellaneous Tree Crops and Groves* : Included all cultivable land which is not included in the area sown, but is put to some agricultural use. Land under groves, forests of timber and full trees, shurbs, bushes etc., which are not included under orchards, are classed under this category.

6. *Culturable Waste Lands* : These include land available for cultivation, whether or not taken up for cultivation or abandoned after a few years for one reason or the other. Such lands may be assessed or unassessed and may be in isolated blocks or within cultivated holdings. Land once cultivated but not cultivate for five years in succession has also been included under this category.

7. *Current Fallows* : These include the cropped areas which are kept fallow during the year.

8. *Other Fallow Lands* : These include all lands, which are taken up for cultivation but are temporarily out of cultivation for a period of not more than five years.

9. *Net Area Sown* : This consists of all the areas being cropped every year or at the reporting year. Area cropped more than once are counted only once.

It has been found from a closer observation of land utilization data as obtained from Governmental agencies that it has a number of discrepancies and distortions. For example, it is but natural to expect the area under the nine land utilisation categories should add up to total geographical area of a spatial unit which was not the case in some instances. Similarly there cannot be any wild fluctuations like sudden increase or abrupt decrease in the area covered by forests since the forests would not come up over a year or two. Even the geographical area of a spatial unit varied in certain instances though there was no change in its boundary.

Land utilisation data for twenty years period from 1961 to 1981 were collected on a year-wise and taluk basis. But for the reasons mentioned above only such data which could be relied upon was made use of. The data which had glaring discrepencies and distortions in it were discarded. The land utilisation data for 1971 and 1981 were found to be more reliable then the other years and hence were taken into consideration. Mandal-wise data for the year 1985 have been used wherever it was found necessary.

Before analysing the distributional pattern of various land-use categories it would be worthwhile to examine the arable and non-arable components of land. The following grouping of land utilisation categories was made:

Arable	None-Arable
Net area sown	Forests
Current fallows	Barren and uncultivable land
Other fallows	Permanent-Pastures
Miscellaneous Tree Crops	Land put to non-agricultural use.
Cultivable Waste Land	

An examination of Table 3.1 reveals that arable lands proportion to the total land in Sriramsagar Command Area is more akin to the figures for Telangana

TABLE 3.1 : **Sriramsagar and Nagarjunasagar Command Areas : Arable and Non-Arable Lands - 1971, 1981 and 1985**

	% Arable Land			% Non-Arable Land		
	1971	1981	1985	1971	1981	1985
Sriramsagar Command Area	56.0	57.5	65.6	44.0	42.5	34.4
Nagarjunasagar Command Area	72.6	73.2		27.4	26.8	N.A.
Telangana	59.6	61.0		40.4	39.0	N.A.
Andhra Pradesh	57.7	58.1		42.3	41.9	N.A.

Computed from the data provided by Bureau of Eco. & Statistics.

and Andhra Pradesh rather than to the figures for Nagarjunasagar Command Area. In Nagarjunasagar Command Area, the proportion of arable land is nearly three-fourths of total land whereas in Sriramsagar Command Area and also in Telangana and Andhra Pradesh, arable land accounts for much less proportion of total land. This only indicates that full impact of irrigation facilities has not yet been felt in Sriramsagar Command Area. As against this, full impact of this factor is evident in Nagarjunasagar Command Area. Relatively low proportion of arable land in Sriramsagar Command Area could also mean either the scope for conversion of non-arable land into arable land by the changes in farming technology etc., is limited or the undulating topography of the area is such that it makes impossible for the water to reach elevated areas. However, proportion of arable land in Sriramsagar Command Area increased to 65.6 per cent of the total area in the year 1985 by which time most of the first phase command area secured irrigation facilities.

If we examine the taluk-wise pattern of arable land (Table 3.2), it is evident that no general trend of increase

TABLE 3.2 : **Sriramsagar Command Area : Arable and Non-Arable Land Taluk-wise - 1971-81**

S.No.	Taluk	Arable land as % of Geographical Area		Non-Arable land as % of Geographical Area	
		1971	1981	1971	1981
1.	Karimnagar	83.2	74.7	16.8	25.3
2.	Metpalli	70.8	60.2	29.2	39.8
3.	Jagtial	52.9	64.8	47.1	35.2
4.	Peddapalli	71.0	51.4	29.0	48.6
5.	Manthani	23.8	27.9	76.2	72.1
6.	Huzurabad	67.9	71.6	32.1	28.4
7.	Nirmal	43.1	50.2	56.9	49.8
8.	Khanapur	11.5	27.7	88.5	72.3
9.	Lakshettipet	25.5	50.4	74.5	49.6
10.	Warangal	44.8	61.3	55.2	38.7
11.	Parkal	74.6	57.5	25.4	42.5
	Command Area	56.0	57.5	44.0	42.5

or decrease in the arable and non-arable components of taluks can be identified. However, the proportion of arable land hiked in the irrigated taluks like Jagtial, Khanapur and Lakshettipet whereas in unirrigated taluk of Parkal, registered decline in its proportion. This could be due to the fact that the entire Command Area is still in a state of transition and hence has not yet fully adopted itself to the changed scenario.

It may be observed from Table 3.3 that, generally speaking, the land utilization pattern of Sriramsagar Command Area resembles, to a certain extent, with that of Telangana and Andhra Pradesh rather than with the land utilization pattern of Nagarjunasagar Command Area. The proportion of forests to the total geographical area has remained more or less the same in both the Command Areas as well as in the State as a whole. So also the categories of barren and culturable waste lands and the land put to non-agricultural use. However, there is a general decline in the area of permanent pastures though this happened on a small scale. The area under tree crops does constitute only a microscopic proportion of the total land. This category is important only in the context of the State. It can safely be foreseen that the area under tree crops would further diminish

TABLE 3.3 : **Sriramsagar and Nagarjunasagar Command Areas : Land Utilisation - 1971-81**

	Year	As percent of Total Geographical Area									
		Forests	Barren & Unculti- vable land	Land put to Non-Agricul- tural Use	Permanent Pasture	Miscella- neous tree crops	Culti- vable. waste	Other Fallow lands	Current Fallow lands	Net Area Sown	Total
Sriramsagar	1971	24.5	7.0	6.6	5.9	0.8	1.6	2.7	13.4	37.5	100
Command Area	1981	25.7	5.7	6.2	4.9	0.5	1.7	2.9	13.2	39.2	100
Nagarjunasagar	1971	8.4	4.9	5.2	8.9	0.7	1.9	7.3	8.9	36.0	100
Command Area	1981	6.7	5.2	6.7	8.2	0.7	2.3	4.9	12.9	52.4	100
Telangana	1971	24.7	5.7	5.3	5.0	0.7	2.3	4.9	12.9	52.4	100
	1981	23.3	5.7	5.6	4.4	0.4	1.8	4.6	11.7	42.5	100
Andhra Pradesh	1971	22.7	8.4	7.5	3.7	0.9	3.7	3.7	8.4	41.0	100
	1981	22.4	8.3	8.0	3.2	0.7	3.1	4.7	9.6	40.0	100

Computed from the data provided by Bureau of Economics and Statistics.

as more and more areas are irrigated and intensive cultivation of land becomes the normal practice in the Command Areas, thereby progressively leading to lessening of area under tree crops.

Expectedly, there are no marked variations in the proportion of barren and uncultivable lands in both the Command Areas. This is in conformity with the figures for the Telangana and Andhra Pradesh. There has been a slight increase in the land put to non-agricultural use in Nagarjunasagar Command Area. However, the same is not applicable in the case of Sriramsagar Command Area. But it can be expected that, with establishments of more and more agro-based and service industries as a direct consequence of agricultural prosperity, more area may be put to non-agricultural use in Sriramsagar Command Area also. On definite pattern that emerges is the slow but steady decline of area under permanent pastures. This may be due to their conversion into agricultural lands after the release of waters.

Both the Command Areas have relatively much less proportion of land under the category 'cultivable waste' when compared to the State as a whole. Even this low component of culturable waste land is being increasingly cropped. More and more of the land which was once considered as uneconomic for agricultural purposes is being brought under plough, made possible by improvement in farming technology and better and judicious application of inputs. The same trend is evident with regard to 'other fallow lands' and 'current fallow lands'. As a result of the above-mentioned factors, there is a net increase in the net area sown in the two Command Areas of Nagarjunasagar and Sriramsagar which is not so in the case of Telangana and Andhra Pradesh as a whole. The increase in net area sown in Command Areas could directly be attributed to release of waters and consequent conversion of fallow lands and waste lands which are hitherto considered uneconomic for cultivation into cultivated lands.

An observation of Fig. 3.1 showing decade variation in net area sown reveals certain interesting patterns. Decrease in net area sown is found in taluks like Sultanabad, Peddapalli, Manthani, Gangadhara, Karimnagar, Bhimdevarapalli and Huzurabad. In Metpalli, Jagtial, Mallial and Warangal net area sown increased. Lakshetti-

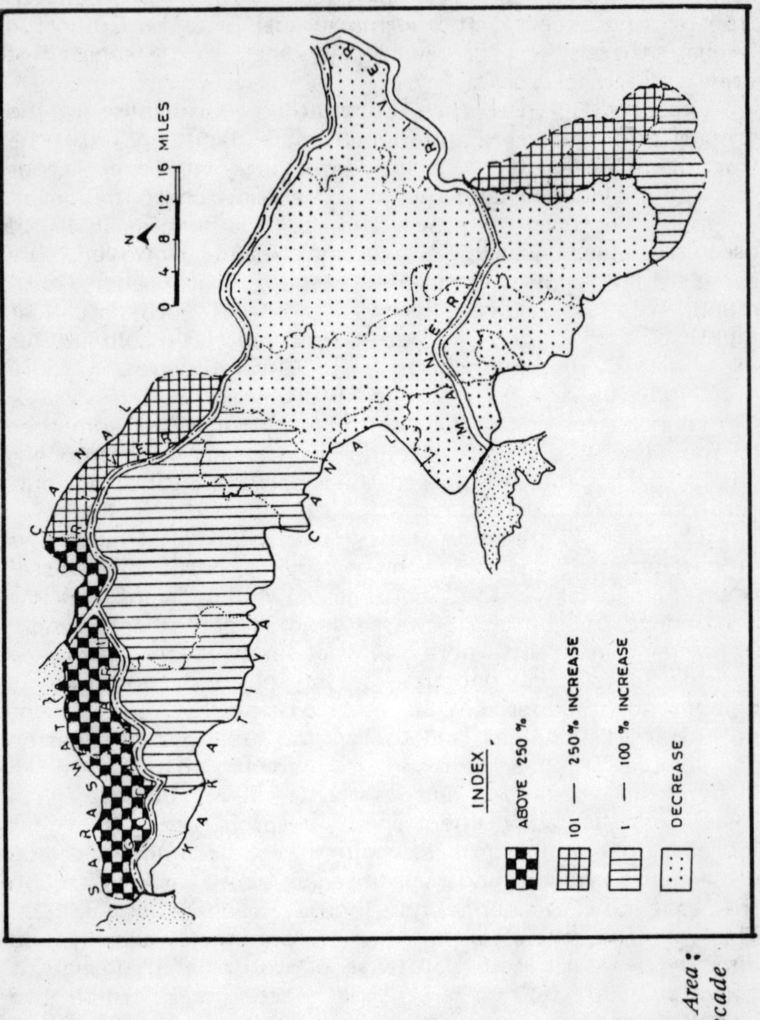

FIG. 3.1: *Sriramsagar Command Area: Net Area Sown Decade Variation 1971-81*

pet, Chityal and Parkal registered greater increase ranging from 101 to 250 per cent whereas Khanapur, Nirmal have registred above 250 per cent increase. Area sown more than once a year (1971-81) (Fig. 3.2) has increased two and a half times in Metpalli, Jagtial, Mallial and other taluks. In Nirmal, Bhimdevarapalli and Huzurabad taluks, decrease was registered. In some taluks increase is found ranging from 1 to 250 per cent.

The net area irrigated as per cent of net area sown increased tremendously from 1971 to 1981 in all the taluks except Parkal where the increase is very marginal. Canal irrigated area as per cent total irrigated area has increased manifold, Metpalli, Jagtial and Lakshettipet received canal water in significant proportions. Whereas Peddapalli and Karimnagar received canal water negligibly. An observation of land use variables gives an increasing trend in some of its indicators. Area sown more than once a year as per cent of net area sown registered general increase except in the taluks of Parkal, Warangal and Nirmal where the decline is noticed. Total cropped area as per cent of geographical area has not registered any change. But proportion of per capita availability of cropped area is slightly increased, whereas, increase is more in the per capita availability of food cropped area. This happened as a direct consequence of the provision of irrigation facility.

Well, Tank and Canal are the predominant sources of irrigation (Fig. 3.3). Canal and Tank irrigation range from 26 to 50 per cent in taluks like Metpalli, Jagtial, Mallial and Sultanabad. Well irrigation is predominant and covering above 75 per cent in Karimnagar, Bhimdevarapalli and Huzurabad taluks; whereas in Manthani taluk tanks account for 70 per cent of irrigated area.

Regional Pattern of Land Utilization

It has already been observed earlier that irrigated part of the Command Area has added more land to its existing arable land during the decade 1971-81. A look at Table 3.4 and Fig. 3.4 reveals certain interesting trends. There was decline in the area covered by forests in the older irrigated taluks like Khanapur and Lakshettipet. In the recently irrigated taluks served by Kakatiya Canal, there is no discernible change. The unirrigated part of the Command Area has little forest occupy more than

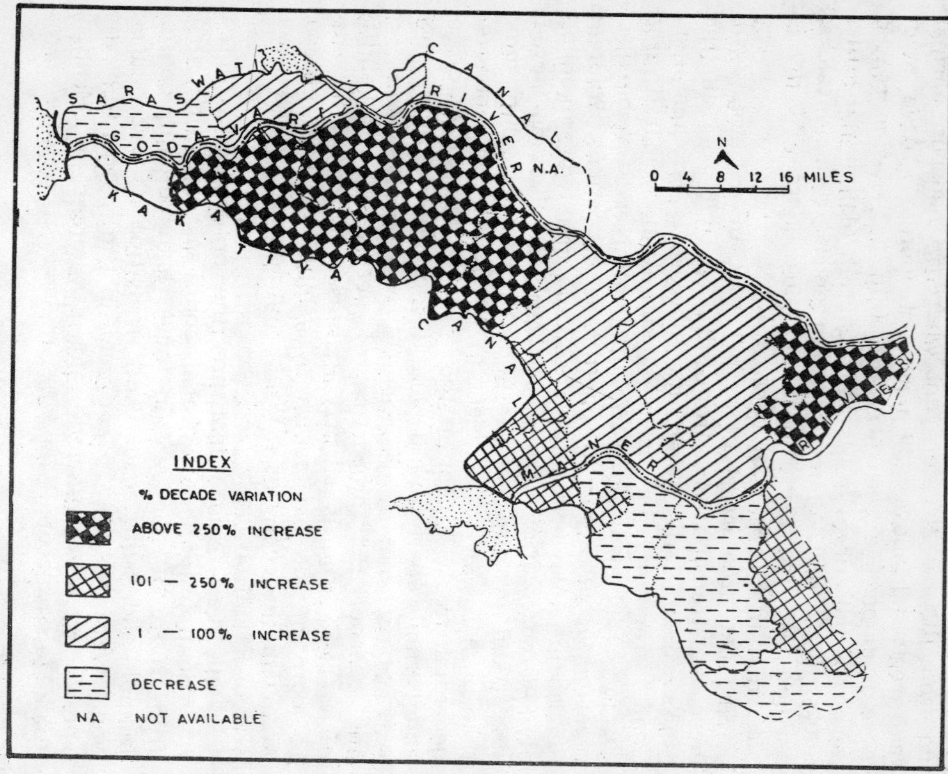

FIG. 3.2:
Sriramsagar Command Area, Sown More Than Once a Year Decade Variation 1971-81

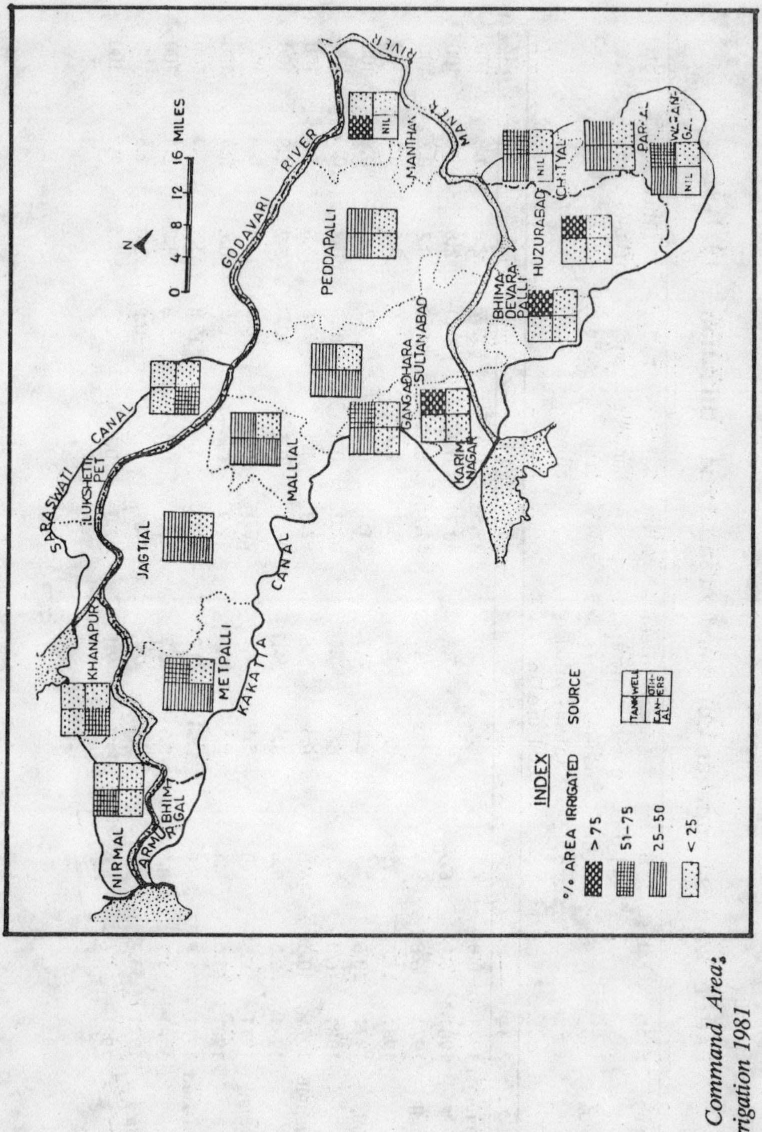

FIG. 3.3: *Sriramsagar Command Area; Sources of Irrigation 1981*

TABLE 3.4 : **Sriramsagar Command Area : Land Utilization - 1971-81**

Sl. No.	Name of Taluk	Year	% Forests	% Barren and uncultivable land	% Land put to non-agricultural use	% Permanent pastures and grazing land	% Net area sown	% Miscellaneous tree crops	% Current fallows	% Other fallows	% culturable waste	Total
1	2	3	4	5	6	7	8	9	10	11	12	13
1.	Karimnagar	1971	0.8	6.4	5.6	4.0	45.5	0.4	32.7	2.8	1.8	100
		1981	6.1	6.3	5.3	7.6	40.7	0.4	29.0	2.8	1.8	100
2.	Metpalli	1971	6.8	9.4	11.3	1.7	47.28	0.02	22.9	0.2	0.2	100
		1981	19.8	9.5	8.4	1.7	50.1	0.2	9.6	0.1	0.2	100
4.	Jagtial	1971	28.6	7.7	6.1	4.74	41.4	0.3	9.7	0.7	0.8	100
		1981	16.8	7.7	6.0	4.7	42.0	0.3	20.9	0.7	0.9	100
4.	Peddapalli	1971	10.3	6.8	6.3	5.6	52.6	1.3	9.6	3.6	3.9	100
		1981	5.3	5.4	3.8	34.1	36.1	0.3	10.0	2.2	2.8	100
5.	Manthani	1971	56.5	6.0	3.1	1.7	20.1	0.	10.5	1.8	0.8	100
		1981	63.4	5.1	2.3	1.3	19.9	0.6	6.4	0.6	0.4	100
6.	Huzurabad	1971	1.8	9.8	11.8	8.7	41.9	1.6	15.8	3.9	4.7	100
		1981	4.7	8.5	8.5	6.7	40.5	1.5	25.8	1.7	2.1	100
7.	Nirmal	1971	33.72	8.2	7.4	7.6	36.5	1.3	0.8	2.0	2.5	100
		1981	26.6	4.4	9.7	9.1	42.0	0.8	2.2	3.3	1.9	100

8.	Khanapur	1971	76.5	6.2	3.8	2.0	3.8	0.5	2.7	4.2	0.2	100
		1981	58.4	4.8	2.1	7.7	21.8	0.1	4.4	0.3	0.4	100
9.	Lakshetti-	1971	67.0	2.7	2.5	2.3	15.8	0.2	8.1	1.2	0.2	100
	pet	1981	58.7	1.9	5.6	3.4	27.3	0.5	14.3	5.9	2.4	100
10.	Warangal	1971	0.5	15.2	17.6	21.9	19.9	2.0	22.9	-	-	100
		1981	-	8.0	13.5	17.1	55.0	0.1	5.2	1.0	0.1	100
11.	Parkal	1971	15.3	3.0	3.2	3.9	9.0	0.6	62.6	1.2	1.2	100
		1981	26.0	5.1	5.3	6.1	5.6	0.9	0.3	10.5	0.2	100

Computed from the data provided by Bureau of Economics & Statistics.

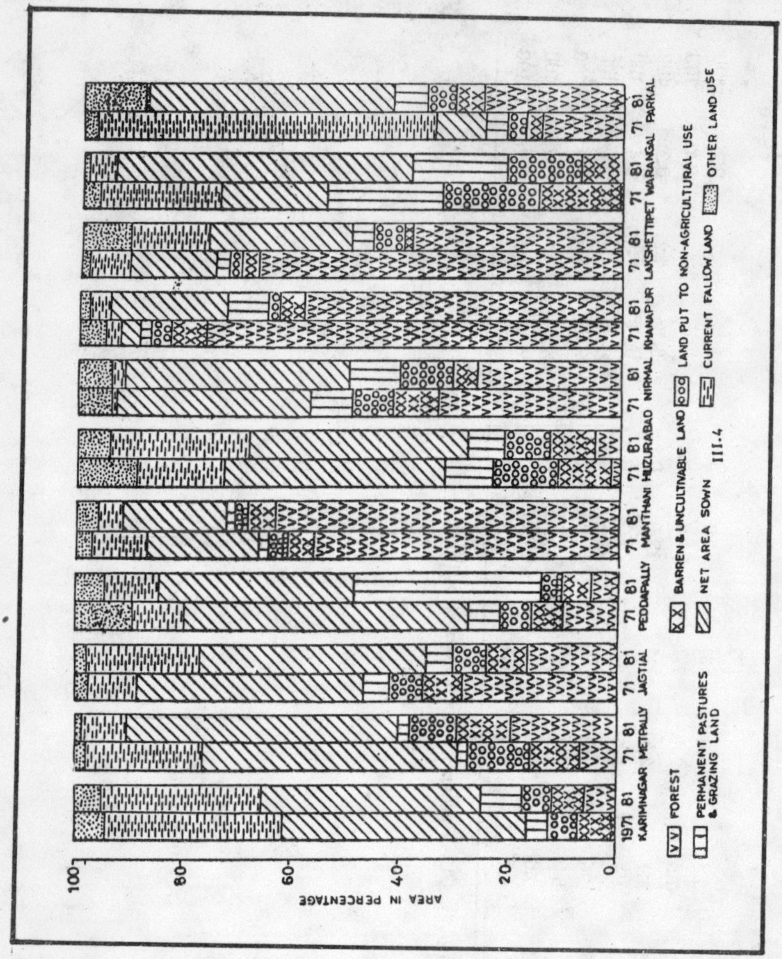

FIG. 3.4:
Sriramsagar Command Area: Land Utilization 1971-81

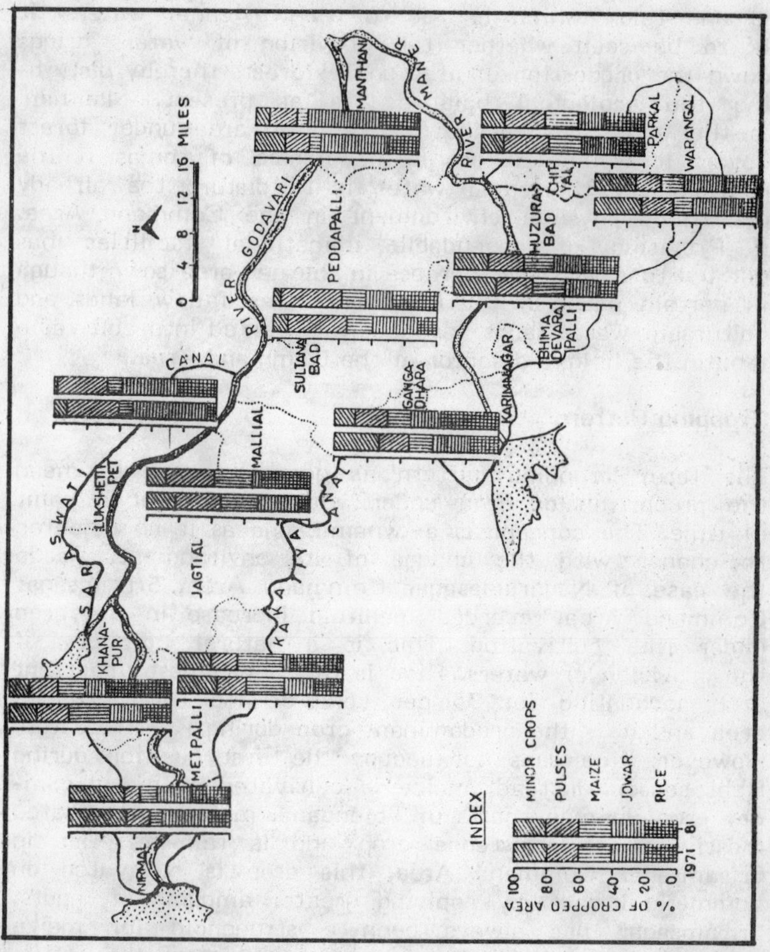

FIG. 3.5: *Sriramsagar Command Area; Cropping Pattern 1971-81*

50 per cent of the total geographical area. Manthani is one taluk which is yet to receive canal waters. It is to be seen whether the provision of waters brings down the proportion of area under forests thereby disturbing the ecological balance. As at present, Manthani is the only taluk having the required area under forest cover. It would not be in the fitness of things if the provision of the canal water is to disturb the already precarious physical environment in the Command Area.

Provision of dependable irrigational facilities has not led to a very big increase in the net area sown though in certain parts of the Command Area fallow lands and culturable waste lands have been converted into cultivable land in the irrigated sector of the Command Area.

Cropping Pattern

The term 'cropping pattern' is generally used to mean the proportion of area under various crops at a point of time. The concept is a dynamic one as it is subjected to change with the change of its environment. As in the case of Nagarjunasagar Command Area, Sriramsagar Command Area recorded manifold increase in the area under rice cultivation. This is a natural corollory of the provision of waters. Rice is the single most important crop accounting for 33 per cent of the total cropped area and it is the predominant crop during Kharif season. However, rice does not occupy the first position during Rabi season. Instead, maize is cultivated as a wet-cum-dry crop. In many parts of Telangana, maize is cultivated mostly as a subsistence crop and is rain-fed. But in Sriramsagar Command Area, this crop is cultivated on commercial basis by applying greater amounts of inputs. Sriramsagar has always been a stronghold for maize cultivation. But after the provision of waters the efficiency with which maize is cultivated has considerably increased.

Study of cropping intensity in an area reveals the degree of cropping in that area, cropping intensity has been calculated for Sriramsagar Command Area for the year 1985. Mandal data have been used for this purpose. The following method has been used for computing cropping intensity:

$$\text{Cropping intensity} = \frac{\text{Gross Cropped Area}}{\text{Net Sown Area}} \times 100$$

Cropping intensity has been found to be high in the areas to which water was released early. The ayacut areas of large tanks which in some cases have been constructed in recent years as part of Sriramsagar irrigation system also have high intensity of cropping. The drier part of the Command Area which does not receive neither sufficient rain nor project waters have very low intensity of cropping.

Rice

The increase in area under rice crop is phenomenal. Normally, it is expected that the area under rice crop would go up with the provision of perennial irrigational facilities. Three-fold increase of rice over a period of ten years (1971-81) has been registered in Sriramsagar Command Area. This is all the more noteworthy since a little more than half of the Command Area was yet to receive canal water (in 1981) and its cropping pattern influenced by the provision of irrigation facilities.

If we look into the taluk-wise distribution of rice cultivation, the irrigated taluks of the Command Area of Sriramsagar, namely Metpalli, Jagtial, Peddapalli, Lakshettipet and Khanapur have greater proportion of cropped area under rice cultivation (Table 3.5). Khanapur ranks first with 41.6 per cent of the total area under rice cultivation in the year 1981. Karimnagar, a part of which receives water from Lower Maner reservoir, also accounts for higher percentage of rice cultivation. In the unirrigated sector of the Command Area, though rice is still the major crop, its proportion to the total cropped area does not compare favourably with the irrigated taluks mentioned above. The unirrigated taluks have less proportion of area under rice cultivation. The proportion of area under rice cultivation varies from taluk to taluk. The unirrigated sector of the Command Area may be having lesser per cent proportion of area under rice cultivation, but it has not experienced any net decline in the area under rice as can be observed from Table 3.6 It can be observed from the table that there has been an increase in the area under rice cultivation in irrigated and as well as in unirrigated taluks of the Command Area. In the irrigated taluks of the Command Area, the per cent decade variation (1971-81) of the area under rice cultivation is high particularly

TABLE 3.5 : **Sriramsagar Command Area : Per cent Area under Crops - 1971-81**

Sl. No.	Name of Taluk	Year	As % of Total Cropped Area											
			Rice	Jowar	Maize	Pluses	Chil-lis	Turme-ric	Cotton	Sesa-mum	Grou-ndnut	Others	Food Crops	Nonfood Crops
1	2	3	4	5	6	7	8	9	10	11	12	13	14	15
1.	Karimnagar	1971	16.84	22.35	14.58	17.93	2.14	0.01	0.58	10.87	13.27	1.43	72.5	27.5
		1981	35.73	17.80	14.13	18.66	2.05	0.01	0.01	1.78	7.12	2.71	89.0	11.0
2.	Metpalli	1971	16.22	19.00	23.16	17.02	1.67	4.36	0.60	13.51	4.00	2.27	82.1	17.9
		1981	32.04	4.26	28.16	14.65	1.12	4.37	0.00	1.56	3.27	10.45	94.8	5.2
3.	Jagtial	1971	7.43	24.4	21.38	27.82	27.93	1.21	0.03	13.38	1.23	0.22	83.1	16.9
		1981	29.53	8.9	23.36	29.51	2.90	2.50	0.00	1.27	0.95	1.08	97.5	2.5
4.	Peddapalli	1971	11.58	26.71	14.77	35.48	1.56	0.04	0.36	3.85	5.12	0.53	88.9	11.1
		1981	28.17	15.26	14.66	35.19	1.96	0.06	0.04	0.08	2.27	0.20	91.3	8.7
5.	Manthani	1971	18.53	42.04	6.32	14.37	0.44	0.00	0.01	4.99	0.33	12.97	81.7	18.3
		1981	25.63	43.43	6.06	16.54	0.39	0.00	0.00	1.15	0.54	7.08	94.4	5.6
6.	Huzurabad	1971	10.24	21.0	20.79	21.03	3.79	0.01	0.34	2.76	19.72	0.32	75.5	24.5
		1981	22.12	16.17	18.87	21.32	6.29	0.00	0.00	0.51	12.56	2.15	85.8	14.2
7.	Lakshettipet	1971	5.04	46.63	8.29	31.61	0.07	0.07	0.04	6.25	0.02	1.98	91.8	8.2
		1981	24.59	24.06	9.65	31.12	0.05	0.02	0.02	0.13	0.00	0.36	99.6	0.6

8. Nirmal	1971	13.77	29.77	7.89	14.97	2.55	1.42	13.08	6.55	3.76	6.22	74.8	25.2
	1981	24.08	21.44	9.55	10.35	2.32	2.21	16.15	2.16	0.17	11.02	70.9	29.1
9. Khanapur	1971	17.45	19.42	25.84	19.84	0.64	1.57	0.41	11.84	0.06	2.95	25.8	14.2
	1981	41.60	11.06	18.66	18.51	1.57	1.70	4.17	1.41	0.01	0.00	93.3	6.7
10. Warangal	1971	8.55	32.59	11.13	20.38	4.68	0.01	0.02	1.28	15.46	5.90	75.6	24.4
	1981	20.07	26.05	15.38	15.0	3.56	0.02	4.63	5.29	8.24	1.26	71.2	18.8
11. Parkal	1971	15.02	23.22	19.97	28.44	6.15	0.00	0.01	0.48	4.88	1.82	93.4	6.6
	1981	24.37	17.01	19.76	24.76	4.53	0.01	0.76	0.42	7.03	1.35	91.1	8.9

Computed from the data provided by Bureau of Economics and Statistics.

TABLE 3.6 : **Sriramsagar Command Area : Per cent Decade Variation in Area Under Principal Crops**

Sl. No.	Taluk	% Decade Variation							
		Rice	Jowar	Maize	All Cereals	Pulses	Total food crops	Total Non-food crops	Total cropped Area
1.	Metpalli	149.8	-71.4	51.0	54.7	6.8	37.6	65.7	19.2
2.	Jagtial	396.2	-36.6	36.4	393.3	32.5	39.7	-82.2	-34.5
3.	Peddapalli	166.6	-35.8	8.8	164.1	10.1	17.3	-70.7	+ 7.5
4.	Karimnagar	138.6	-10.5	9.0	136.8	17.0	34.6	34.6	9.7
5.	Manthani	38.6	1.1	-4.3	37.7	15.4	15.4	15.4	-0.4
6.	Huzurabad	113.7	-27.4	-10.3	110.8	0.3	6.9	6.9	-5.8
7.	Warangal	9.8	-64.7	-33.6	7.9	-65.8	-57.9	-57.8	-57.8
8.	Parkal	97.6	-10.8	20.6	104.4	6.1	18.9	18.9	21.8
9.	Lakshettipet	375.7	-28.7	13.6	363.9	-3.9	5.6	-94.0	-2.7
10.	Nirmal	106.5	-14.9	42.9	101.6	-18.4	17.7	-36.2	18.2
11.	Khanapur	207.0	-26.6	-6.9	296.9	20.3	40.2	-39.3	28.9

Computed from data provided by Bureau of Economics and Statistics.

so in the taluks of the Jagtial, Lakshettipet and Khanapur. The lowest value of the per cent decade variation in area under rice cultivation is found in the case of Warangal followed by Manthani.

Jowar

The per cent proprotion of area under Jowar cultivation declined from 27.8 per cent in 1971 to 19.2 per cent in 1981 in the Command Area as a whole. This decline is more or less in conformity with the trend in other parts of the State. However, if we examine taluk-wise per cent proportion of area under rice cultivation, some interesting patterns emerge (Table 3.5). There has been a sharp decrease in the per cent proportion area under Jowar cultivation in the irrigated sector of the Command Area. For example, Metpalli had 19.00 per cent of its total cropped area under Jowar crop in 1971 which dropped to 4.26 per cent in 1981. The same trend holds good in the case of Jagtial and Peddapalli also. However, Jowar crop's proportion in the cropped area did not come down sharply in Lakshettipet, Nirmal and Khanapur taluks. Similarly Jowar continues to be the major millet crop in the irrigated sector of the Command Area. If we examine the per cent decade variation in the area under Jowar crop, it can be seen that only one taluk namely Manthani increased its area under Jowar very marginally. All other taluks of the Command Area registered a net decrease during the decade 1971-81 (Table 3.6). The highest negative per cent decade variations are observed in the irrigated sector of the Command Area with Metpalli ranking first (-71.4%).

Maize

Superficially, the increase in the per cent proportion of area under maize cultivation to total cropped area is not very high (14.8% in 1971 and 17.7% in 1981). Similarly maize ranks only as the third most important of all crops in the Command Area as a whole. It ranks next to Jowar. However, the present importance of maize crop and its future potential in the Sriramsagar Command Area cannot be minimised. In fact maize is a crop which is very characteristic of the Command Area. In no other part of the State maize cultivation is more important than in this Command Area.

A closer look at Table 3.5 reveals that decrease in Jowar's component of the total cropped area is in line with the increase in maize crop contribution to total cropped area. We can safely deduce from this, that Jowar which cannot be cultivated on a very intensive scale and which is mainly a dry crop, is giving way to maize cultivation for which possibilities for intensive cultivation exist. The general trend of increase in maize crop's contribution to total cropped area is however does not apply in the case of northern taluks namely Lakshettipet, Nirmal and Khanapur. It may be assumed here, that a decline in the Jowar's proportion of the total cropped area is compensated in these taluks by increase in proportion of area under rice cultivation, and not by maize as happened in the case of rest of the Command Area. The unirrigated taluks of the Sriramsagar Command Area do not exhibit any glaring pattern about maize cultivation, though maize continues to be the third most important crop.

If we examine the net increase, the area under maize cultivation during the last decade registered a positive variation of 51 per cent in Metpalli taluk (Table 3.6). This is also the taluk which registered the highest negative decade variation in Jowar cropped area during the same period. All other taluks of irrigated sector of the Command Area also registered positive decade variations in the area under maize cultivation except the taluk of Khanapur. There was a general fall in the area under maize cultivation in the unirrigated taluks.

As mentioned already, maize cultivation is on the ascendency both in quantitative and qualitative terms. With the extension of irrigation facilities the importance of maize crop will further go up. In course of time, it may replace Jowar as the most important millet crop. There are very good chances of its emerging as a good substitute crop in place of rice specially during those years when the Sriramsagar reservoir does not receive its full quota of water. One advantage of maize cultivation is its relatively less dependence on water, labour and other inputs. Further, the demand for maize is increasing steadily in the industrial sector. It constitutes the most important part of the feed required for poultry farming whose activity is rapidly increasing in the State.

Minor Crops

Provision of irrigation facility is felt on the cultivation of principal and minor crops. The proportion of principal crops has increased from 77.5 per cent in 1971 to 94.2 per cent in 1981. In direct consequence to this, the proportion of minor crops has drastically come down. The minor crops comprise of Chillies, Turmeric, Cotton, Sesamum, Groundnut etc. The proportion of individual crops is considerably low when compared with principal crops like rice, jowar, maize and pulses.

A closer examination of Table 3.5 reveals the following pattern with regard to minor crops. Although the proportion of Chillies is on the whole low, its economic importance is more and it has maintained stability with little variations from 1971 to 1981. Its proportion ranges from a mere decimal value in some taluks to about 6.29 in Huzurabad taluk. Another cash crop Turmeric is highly localised and is confined to the taluks like Metpalli, Jagtial, Nirmal and Khanapur. Area under turmeric has increased only slightly. Cotton is grown mostly in Nirmal taluk and it has registered a marginal increase in taluks like Nirmal, Khanapur and Warangal. In other taluks cotton is grown in a negligible area. Sesamum occupies high proportion among the minor crops in 1971. Its share has sharply declined and had its effect on the proportion of minor crops as a whole. Groundnut occupies a considrable proportion both in 1971 and 1981. It is grown in large areas in Karimnagar, Huzurabad, Warangal and Parkal. In general, the share of groundnut crop has decreased from 1971 to 1981.

Impact of Irrigation on Cropping Pattern

That the provision of assured irrigation facilities has its impact on cropping pattern and intensity is widely accepted and is a proved hypothesis. As has been pointed out earlier in this narration, the cropping pattern in Sriramsagar Command Area has undergone many changes. But it is essential to establish in comparable quantitative terms what the cropping pattern was before and after the release of waters in the irrigated taluks of the Command Area. This is specially so if the water is released in a phased manner. To cite an instance, the cropping pattern of Metpalli for which water was released more

TABLE 3.7 : Sriramsagar Command Area : Release of Waters and Change in Cropping Pattern (Base Value 100)

Sl. No.	Taluk	9 Years before	6 Years before	3 Years before	Year Water released	3 Years after	6 Years after	9 Years after
1	2	3	4	5	6	7	8	9
					Rice			
1.	Metpalli	102	80	113	100	162	163	180
2.	Jagtial	61	26	27	100	133	124	-
3.	Peddapalli	43	55	93	100	127	-	-
					Jowar			
1.	Metpalli	101	87	97	100	81	64	32
2.	Jagtial	109	107	113	100	101	62	-
3.	Peddapalli	104	110	90	100	66	-	-
					Maize			
1.	Metpalli	78	80	93	100	108	119	138
2.	Jagtial	76	79	88	100	105	117	-
3.	Peddapalli	70	95	112	100	104	-	-

				All Cereals			
1.	Metpalli	104	80	113	100	163	164	180
2.	Jagtial	60	26	27	100	133	124	-
3.	Peddapalli	43	55	93	100	127	-	-
				Food Crops			
1.	Metpalli	105	91	101	100	109	105	108
2.	Jagtial	84	77	81	100	111	105	-
3.	Peddapalli	79	89	94	100	101	-	-
				Non-Food Crops			
1.	Metpalli	194	95	110	100	92	19	30
2.	Jagtial	239	207	227	100	73	33	-
3.	Peddapalli	99	198	141	100	51	-	-

Note : Year Release of Water (Base Value 100), Metpalli - 1970-71; Jagtial - 1974-75; Peddapalli - 1977-78

Computed from Bureau of Economics & Statistics data.

than ten years back is not exactly comparable with that of Peddapalli which received water about 3 years back.

Conventional methods of comparing the cropping patterns existing at different periods of release of waters cannot be adopted in such cases. In order to overcome this methodological problem, recourse has been taken to rearrange the data with the year of release of waters as the base with a value of 100 and the deviations from this base for 3, 6, 9 years before and after the release of waters, for the three irrigated taluks served by Kakatiya Canal. The figures thus computed are shown in Table 4.4.

The figures confirm the observation made earlier in this chapter. One interesting factor which emerges from an analysis of table is that cropping pattern fluctuate greatly before the release of waters. Once the waters are released the increase or decrease of area under crops/crop groups stabilises itself. The period after the release of water in the irrigated taluks of Sriramsagar Command Area is characterised by a steady increase in rice and maize cultivation. Food crop's proportion is on the ascendency and that of non-food crops on the descent. In Metpalli, rice crop's position was unstable before the year of release of waters and ranged between '80' and '113'. However, there was a steady increase during the next nine years. The present index value being '180'. In the case of maize, the instability before the release of waters was not marked and the area under maize crop has increased from index value of '78' nine years before the release of water to '100' in the year of release of waters, then to '138' nine years thereafter. The decline in the case of jowar was not very marked before the release of water, but it was a case of steady decline to an index value of '32' nine years after the provision of irrigation facilities. The trend that has been described for Metpalli taluk holds good in the case of other taluks too.

Index of Concentration of Maize and Rice Crops

In order to probe into the spatial aspect of the distribution of two important crops of the Command Area, the index of concentration of rice and maize has been calculated on mandal basis for the year 1985.

$$\text{Index of Concentration} = \frac{\dfrac{c_i}{c_t}}{\dfrac{C_i}{C_t}}$$

where,

c_i = area under the crop in the Mandal

c_t = total cropped area in the Mandal

C_i = area under the crop in Command Area

C_t = total cropped area in Command Area.

The index of concentration is high in irrigated sector of the command area. The high concentration of the rice in the irrigated sector is understandable. Maize which was hitherto used to be grown as a dry crop, has now emerged as an important wet-cum-dry crop. Maize is now grown as a commercial crop.

Agricultural Efficiency

Agricultural efficiency has been the subject of study by many geographers in India and abroad. Various methods have been employed by taking different variables to measure the agricultural efficiency. Ali Mohammed (1978) in his authoritative work "Studies in Agricultural Geography" surveys various methods of measuring agricultural efficiency. According to him J.L. Buck (1967) evolved a method with output expressed in terms of grain equivalents. He considered all foodgrains to be equivalent in food values and took kilogram as unit. All other products were converted into grain equivalents on the basis of the amount of the most usually consumed grain of the locality that the product could buy. E. de Varies (1967) expresses the output in terms of 'milled rice equivalents' for head of population according to the local market price to each grain, whereas Buck (1967) has considered all grains as equal. Ali Mohammad (1978, p. 74) criticises saying that "this method will not be applicable especially in the multi-crop regions as there may not be a single diet much more prevalent in dietary habit of people". Clark and Hswell (1967) further modified E. de Varies method and expressed the agricultural ouput in terms of wheat equivalent kilograms available per person. Kendall (1939) devised a system of ranking co-efficients to measure the agricultural efficiency. Acre-yield of crops as output per unit area has been considered. He took the acre-yield of ten leading crops for 48 administrative countries and for each of ten crops he ranked the countries according to yield. The sum of ranks occupied by a country was divided by the number of crops to obtain the average rank of the country (Ali Mohammed, 1978). Dudely Stamp (1960) applied the technique for the countries of the world and placed them in order of output per acre for each crop. The places occupied by each country in respect to the selected crops were then averaged and from these averages the ranking co-efficient of agricultural efficiency of each country was obtained.

Shafi (1960) applied the method for the first time in India. He considered the acre yields of 8 crops to measure agricultural efficiency in Uttar Pradesh. Ali Mohammad (1978, p. 75) criticises that "this method seems to have vital and inherent weakness, which makes it some what insensitive as a measure of agricultural efficiency. This weakness arises from the neglect of the areal strength of crops for which acre yields are taken to calculate the ranking co-efficient". He adds that "a crop which occupies a negligible portion of crop land would contribute almost nothing to agricultural efficiency of the serial unit even though it may have acre-yield".

Deshpande and Sapre (1954) have tried to eliminate the weakness in Kendall's method modifying it by taking a weighted average of the ranks. Weights for ranks of various crops being proportional to the percentage of crop land under respective crops. However, this method is not effective than Kendall's rank co-efficient because the magnitude of difference in acre yields is not concealed as mentioned by Ali Mohammad (1978). Khusro, as summarised by Singh (1976), proposed a method determining the output in relation to input or input-output ratio and profitability of farming measured in terms of the return for the sum total of human effort or paid out cost in relation to the output.

Bhatia (1965) proposed a method based on the weightage of acre-yield in respect of the proportionate share of each crop to the total cropped acreage of any serial unit. He measured the agricultural efficiency by evolving the following formula :

$$I_i = \frac{\frac{P}{J} = IE_{ij} \quad C_{ij}}{\frac{P}{J} = I \quad C_{ij}}$$

$$I_j = \frac{Y_{ij} \quad x \quad 100}{YJ}$$

Where

I_i = Index of Crop Productivity in *i*th District.

Y_{ij} = Yield per acre of *j*th crop in *i*th District.

Y_j = Yield per acre of the *j*th crop in the region.

C_{ij} = Proportionate share of land under *j*th crop in the total cropped area of the *i*th district.

Singh (1976, p. 318) observes that, "calculating the index number of agricultural efficiency by expressing per unit area carrying capacity in terms of population of the component enumeration unit as a percentage of the per unit area carrying capacity for the entire region involves lengthy and complicated calculations". He adds that, "Evalution of various approaches to measure the agricultural performances reveals that most of them suffer from some noted weaknesses. Keeping this in view, a new technique of measuring the level of agricultural production is adopted, *i.e.* the crop yield and concentration indices, ranking co-efficient". In order to measure the Index, he used the average foodcrop yields and proportion of these crops in the total harvested area. The relative crop yield and concentration indices arranged into ranking order and computed into average ranking co-efficient. He expressed the formula as under:

$$Y_i = \frac{Y_{a_e}}{Y_{a_r}} \times 100$$

Where

Y_i is the crop yield index,

Y_{a_e} is the average yield per hectare of crop, a in the component enumeration unit and

Y_{a_r} is the average yield of the crop a in the entire region.

$$C_i = \frac{P_{a_e}}{P_{a_r}} \times 100$$

Where

C_i. is the crop concentration index.

P_{a_e} is the percentage strength of the crop a in the total harvested area in an enumeration unit.

P_{a_r} is the percentage strength of the crop a in the total harvested area in the entire region or state.

Yield and concentration indices ranked for individual crops are added and thereafter divided by '2'.

$$\text{Crop Yield and concentration Indices Ranking Co-efficient for crop 'a'} = \frac{\text{Crop yield index Ranking of Crop 'a'} + \text{Crop Concentration index Ranking of crop 'a'}}{2}$$

The lower the ranking co-efficient, the higher is the level of agricultural production and *vice versa*.

Singh (1976) has purely relied upon food crop yields and proportion of these crops in measuring the agricultural efficiency. The other indicators like agricultural innovations, irrigation, demographic etc. have not been taken into account.

Method used in the Present Study

Provision of assured irrigation facilities as a rule leads is increased level of agricultural efficiency development and agricultural efficiency. The aim of foregoing discussion is to quantitatively identify the level of agricultural efficiency in the sixteen taluks of Sriramsagar Command Area. Here, it is necessary to distinguish between agricultural efficiency and agricultural productivity. The index of agricultural productivity is more concerned with measuring the production and the area of different crops. As against this, the Index of Agricultural Efficiency is comprehensively concerned with the efficiency or otherwise with which agricultural activity in a given area is carried on. Agricultural efficiency cannot be measured by just considering a few indicators like yield area, production etc. If the Agricultural Efficiency Index is really to be made meaningful it should encompasse a wider horizon of indicators.

Two sets of problems arise in measuring the Agricultural Efficiency. First problem relates to the selection of meaningful indicators of agricultural efficiency. This, in turn depends on the limitations imposed by data availability. Seldom do we find the right type of data. Second problem is with regard to determination of the relative importance of the indicators, since, all the indicators do not promote efficiency of agriculture equally. These problems have been dealt with as mentioned below.

Sixteen indicators of Agricultural Efficiency have been selected after a careful study of their relative importance in increasing efficiency. The '16' indicators have been classed into five groups. The groups and the indicators which fall in these groups are given below:

1. *Irragation and Rainfall Indicators*

1. Net irrigated area as per cent of net area sown.
2. Canal irrigated area as per cent of total irrigated area.
3. Mean annual rainfall

2. ***Yield Indicators***

1. Yield of rice
2. Yield of maize
3. Area under high yielding rice as per cent of total rice area.
4. Area under high yielding maize as per cent of total maize area.

3. ***Land Use Indicators***

1. Area sown more than once a year as per cent of net area sown.
2. Total cropped area as per cent of geographical area.
3. Per capita availability of cropped area.
4. Per capita availability of food cropped area.
5. Culturable waste as per cent of geographical area.

4. ***Agricultural Innovations***

1. Population power tillers, tractors ratio.
2. Electrified fillages as per cent of total villages.

5. ***Demographic Indicators***

1. Population-literacy ratio
2. Per cent owner cultivators.

Irrigation is generally recognised as a factor in increasing the agricultural efficiency. But canal irrigation is much more dependable and leads to stable agricultural activity. Therefore, the three indicators given in group I have not been considered as having equal degree of importance in determining agricultural efficiency. Canal irrigated area has been given more weight than net irrigated area. Since all the taluks of Command Area are not presently irrigated by canals, mean annual rainfall of the taluk is also considered as an indicator although it is assigned less weight.

Second group of indicators pertain to yield and the area under high yielding varieties. Yield is certainly reflective of agricultural efficiency. Yields of the two principal crops namely rice and maize have been taken into consideration in addition to the proportion of area under high yielding varieties of these two crops.

It can be safely assumed that an area which is sown more than once a year is more efficient agriculturally. It also means that land is put to intensive use if the area is sown more than once a year. Similarly total cropped area as proportion of the geographical area indicates the degree of use to which land area in a spatial unit is put to. Since cropped area alone does not give a full picture if population is not correlated to it, the per capita availability of cropped area and also food cropped area have been considered as indicators. In contrast to cropped area, the proportion of culturable waste to total geographical area denotes that land is not put to maximum use. Greater the area of culturable waste in a spatial unit lesser is the efficiency of agriculture and *vice versa*.

Among the agricultural innovations, power tillers and tractors greatly contribute to carry on an agricultural activity on more efficient lines. But the absolute number of power tillers and tractors may not mean anything if we do not take into consideration the size of population for which they are available. Therefore, the population-power tiller, tractors ratio was computed. Another infrastructure which promotes agricultural efficiency is the availability of electrical power. Electrified villages as per cent of total villages has therefore been taken as one of the indicators of agricultural efficiency.

A literate farmer is normally a more capable and better farmer, because he keeps himself aware of the new agricultural innovations. Because of his amenability to changes, he adopts new techniques of agriculture which naturally results in more agricultural efficiency. Similarly, it has generally been found from our knowledge of local conditions in Command Area that an owner cultivator takes more interest in his land than a cultivator who tills the land as a tenant.

Having selected the indicators after a careful examination of their importance as promoters of agricultural efficiency, a suitable method has to be evolved to amalgamate all these varied indicators on a quantitative basis. After an evaluation of the various techniques used by others, it has been found that no single method is comprehensive enough to give a total picture. The present method is comprehensive enough to give a total picture. The method adopted in this study is detailed below.

Quantification of Agricultural Efficiency

Various of the sixteen indicators greatly differ in their units, values and their relative importance. The purpose here is to convert the values in each of the sixteen indicators (or distributions in statistical terminology) into comparable unit values. This can best be done by computation of standard scores. Since standard scores are equal units of measurement and since their size is the same from distribution to distribution, they become an useful tool in the present study. The formula for the standard score is as follows :

$$\text{Standard Score } (z) = \frac{(X - \overline{X})}{S} = \frac{x}{S}$$

Where

X = Any raw score or unit of measurement in a distribution

$\overline{X}$ = Mean value for the distribution (indicators)

S = Standard Deviation of the distribution or indicator:

$$x = (X - \overline{X})$$

The mean is subtracted from each of the values within a distribution, and then the deviation from mean $(X - \overline{X})$ or x is divided by the standard deviation to change the deviation values into standard scores (z). When we change raw scores to standard scores, we are expressing them in standard deviation units. These standard scores tell us how many standard deviation units any given score deviated from the mean. Since three Standard Deviations on either side of the mean include practically all the cases, it follows that the highest z score usually encountered is +3.00 and the lowest –3.00.

The standard scores are equal units of measurement and hence can be manipulated mathematically. It should be noted here that changing a distribution of scores to z scores does not change the shape of the original distribution of scores. Since z scores are expressed in decimals and since some of them are negative, they are rather cumbursome to handle. In order to overcome this difficulty linear transformation of *z* values are, therefore, made here. Such transformations consists of making the scale larger, so that negative scores are eliminated and decimals are done away with. Transformed scores have been obtained from the following equation:

Transformed Standard Score = Z (New Standard Deviation) + New Mean

The transformed standard score thus obtained for each of the indicators are given in Appendices 3.1 and 3.2. Since all the indicators are not equally important, it is necessary to assign weights for them. The following weighted and transformed standard scores have been obtained. Cumulative weighted and transformed standard scores for the taluks were computed and the mean of these values calculated for each of the taluks (Tables 3.8 and 3.9). The values thus obtained have been taken into consideration for classifying the taluks with regard to their level of agricultural efficiency. The index of Agricultural Efficiency for a taluk can finally be expressed in the following equation form :

$$IAE = \left(T_{z_1} + W_1\right) + \left(T_{z_2} + W_2\right) \ldots \left(T_{z_n} + W_n\right) \quad N$$

Where

IAE = Index of Agricultural Efficiency for the Taluk

T_{z_1} = Transformed standard score of Indicator 1

W_1 = Weight assigned for indicator 1

N = Number of indicators.

Weights assigned for indicators

1. *Irrigation indicators*

1. Net irrigated area as per cent of Net Area Sown '10'
2. Canal Irrigated area as per cent total Irrigated Area '15'
3. Rainfall '5'

TABLE 3.8 : **Sriramsagar Command Area : Weighted Standard Scores of Agricultural Efficiency Variables - 1971**

S.No.	Taluk	Weighted Standard Scores						
		Net Irrigated Area as % net area shown	Canal irrigated area as % total irrigated area	Mean annual rainfall	Area under high yielding rice as % total rice area	Area under high yielding maize as % total maize area	Area shown more than once a year % of net area sown	Total cropped area % of G. area
1	2	3	4	5	6	7	8	9
1.	Karimnagar	67	54	36	62	75	46	55
2.	Metpalli	60	56	57	41	51	48	57
3.	Jagtial	48	51	56	66	54	47	52
4.	Peddapalli	56	50	57	69	52	52	66
5.	Manthani	51	52	58	41	56	46	36
6.	Huzurabad	74	52	38	53	71	58	59
7.	Lakshettipet	49	62	69	54	44	54	43
8.	Khanapur	49	86	67	47	44	52	37
9.	Nirmal	41	50	65	43	44	49	49
10.	Parkal	68	53	48	59	48	68	58
11.	Warangal	41	52	55	69	65	81	67

Sl. No.	Taluk	Weighted Standard Scores								
		Per Capita availability of cropped area	Per Capita availability of food cropped area	Culturable waste as % of G.area	Population till-Tractors	Electrified village as % of tot.vill.	Literacy Ratio	% owner cultivators	Cumulative weighted standard score.	Average weighted standard score.
1	2	10	11	12	13	14	15	16	17	18
1.	Karimnagar	41	39	50	60	65	66	58	774	55
2.	Metpalli	44	44	43	52	39	46	62	700	50
3.	Jagtial	49	49	47	51	43	49	60	722	52
4.	Peddapalli	57	59	36	66	54	50	24	748	53
5.	Manthani	72	69	47	40	39	43	51	701	50
6.	Huzurabad	44	41	30	45	66	51	59	741	53
7.	Lakshettipet	61	68	45	45	56	54	44	732	53
8.	Khanapur	51	54	44	42	62	44	54	736	52
9.	Nirmal	61	54	43	75	50	52	55	730	52
10.	Parkal	59	64	42	52	42	43	51	863	55
11.	Warangal	39	40	40	46	61	78	58	815	58

TABLE 3.9 : **Sriramsagar Command Area : Weighted Standard Scores of Agricultural Efficiency Variables - 1981**

Sl. No.	Taluk	Net Irrigated Area as % net area shown	Canal Irrigated area as % total irrigated area	Mean annual rainfall	Yield of Rice	Yield of Maize	Area under high yielding rice as % total rice area	Area under high yielding maize as % of total maize area	Area sown more than once a year % of net area sown
1	2	3	4	5	6	7	8	9	10
1.	Metpalli	78	66	57	60	44	33	80	55
2.	Jagtial	67	67	56	55	57	53	60	57
3.	Mallial	45	66	56	62	48	50	59	64
4.	Sultanabad	52	62	57	57	53	54	49	78
5.	Peddapalli	58	51	57	50	62	70	43	47
6.	Manthani	52	48	58	81	49	32	41	46
7.	Gangadhara	55	52	38	46	43	55	61	58
8.	Karimnagar	66	49	36	44	45	48	49	58
9.	Huzurabad	53	49	38	47	45	53	48	51
10.	B.D'palli	47	48	36	56	39	57	48	67
11.	Chityal	62	48	48	58	43	64	56	57
12.	Parkal	57	51	48	49	58	47	45	27
13.	Warangal Rural	62	48	55	42	47	55	44	50
14.	Nirmal	35	58	62	43	78	51	48	41
15.	Khanapur	67	83	67	34	59	73	65	55
16.	Lakshettipet	48	82	69	50	69	55	42	40

Sl. No.	Taluk	Total cropped area % of G. A.	Per Capita availability of cropped area	Per Capita availability of food cropped area	Culturable waste as % of G. area	Population/ till- Tractors	Electrified village as % of tot.vill.	Literacy Ratio	% owner cultivators	Cumulative weighted standard score	Average weighted standard score
1	2	11	12	13	14	15	16	17	18	19	20
1.	Metpalli	58	53	53	62	61	9	49	76	933	58
2.	Jagtial	50	51	56	56	67	31	54	54	894	56
3.	Mallial	55	57	59	62	62	65	36	42	873	55
4.	Sultanabad	65	62	64	42	64	62	45	47	913	57
5.	Peddapalli	57	55	58	21	63	51	61	56	799	50
6.	Manthani	42	70	71	57	50	50	47	37	83.1	52
7.	Gangadhara	58	44	47	48	48	62	53	44	812	51
8.	Karimnagar	56	43	42	46	57	60	85	43	827	52
9.	Huzurabad	64	48	47	50	49	61	52	51	827	52
10.	B. D'palli	40	27	25	49	47	61	50	51	7548	47
11.	Chityal	47	57	57	62	43	57	40	42	848	47
12.	Parkal	68	55	54	62	46	61	50	62	890	56
13.	Warangal Rural	70	48	42	69	59	58	53	56	58	54
14.	Nirmal	48	67	53	47	61	57	50	54	853	53
15.	Khanapur	38	49	50	61	65	50	44	40	901	56
16.	Lakshettipet	38	57	70	48	65	45	57	71	906	57

2. *Yield Indicators*

1. Yield of Rice '5'
2. Yield of Maize '5'
3. Area under high yielding rice as per cent of total rice area '5'
4. Area under high yielding Maize as per cent of total Maize area. '5'

3. *Land Use Indicators*

1. Area sown more than once a year as per cent of net area sown '10'
2. Total cropped area as per cent of geographical area '5'
3. Per capita availability of cropped area '5'
4. Per capita availability of food croped area '5'
5. Culturable waste as per cent geographical area '5'

4. *Agricultural Innovations*

1. Population/power tillers, tractors ratio '5'
2. Electrified villages as per cent total villages '5'

5. *Demographic Indicators*

1. Literacy Ratio '5'
2. Per cent owner cultivators '5'

All the indicators mentioned above have been used for the year 1981. As the yields were not available for the year 1971, the weightage was doubled for high yielding crops so as to equate the weightage of yield indicators with 1981.

It is observed from the analysis that there has been a tremendous impact of irrigation on agricultural efficiency after the provision of irrigation. The taluks which were not provided water in 1971 ranged from low to moderate regions of agricultural efficiency, have risen from moderate to high after provision of water. The other taluks which are yet to receive water have fallen from 'high' to 'moderate' regions of agricultural efficiency (Figs. 3.6 and 3.7). The present agricultural effici-

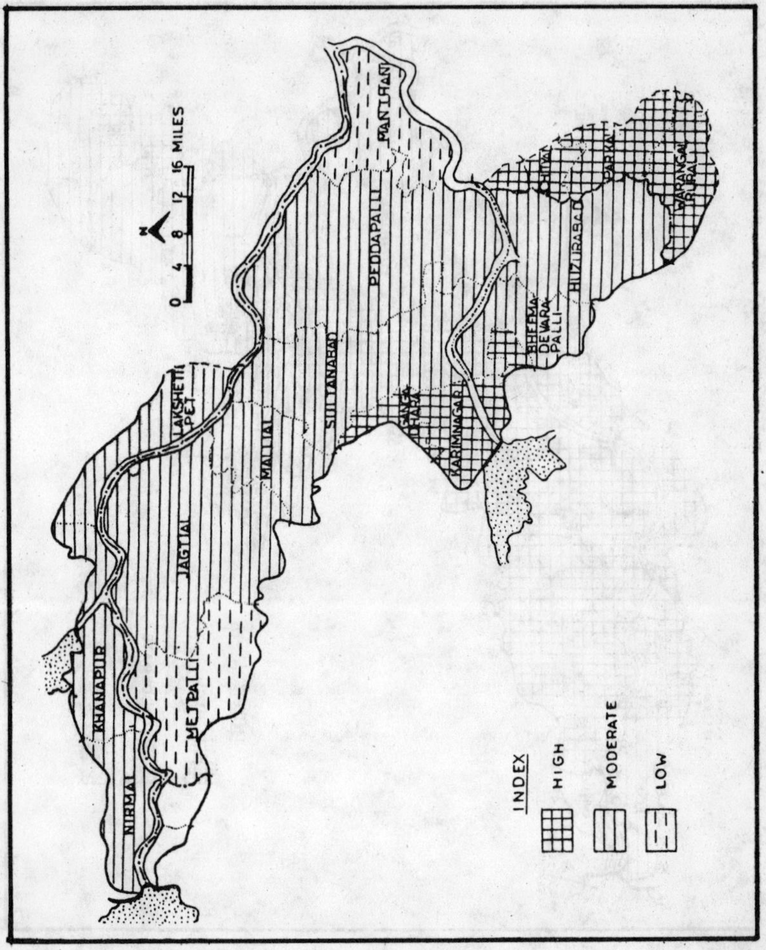

FIG. 3.6:
Sriramsagar Command Area: Agricultural Efficiency 1971

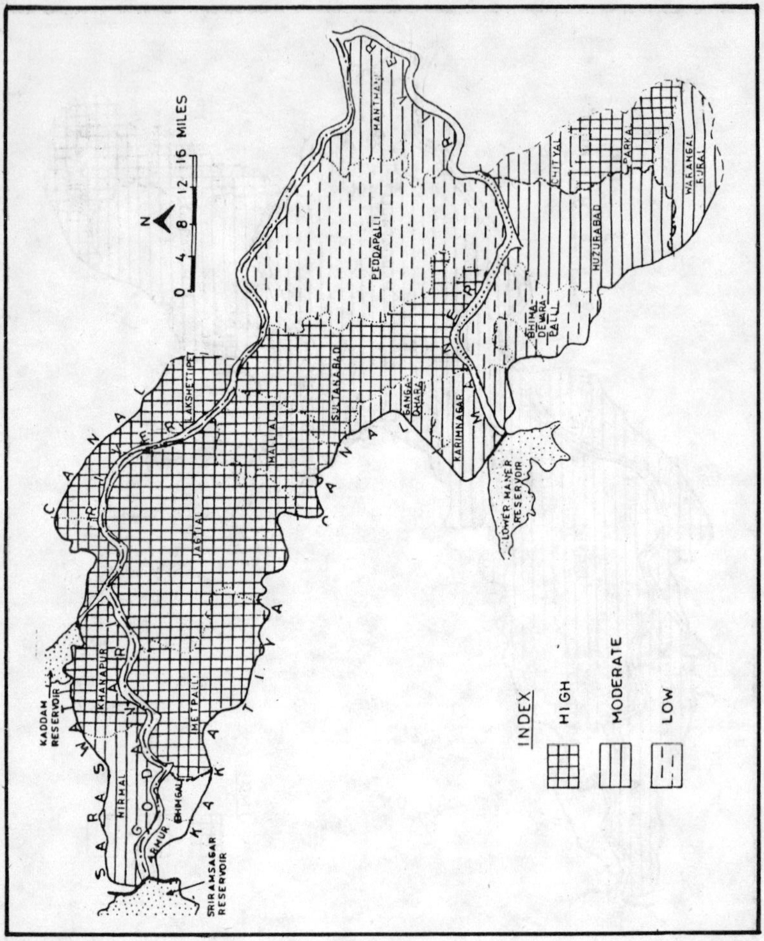

FIG. 3.7: *Sriramsagar Command Area: Agricultural Efficiency 1981*

ency pattern (1981) of the sixteen taluks of Command Area have been categorised (Fig. 3.4) into three Agricultural Efficiency Regions as follows:

1. High agricultural efficiency region.
2. Medium agricultural efficiency region.
3. Low agricultural efficiency region.

High Agricultural Efficiency Region

This region comprises of seven taluks namely Metpalli, Jagtial, Mallial, Sultanabad, Parkal, Khanapur and Lakshettipet. All these taluks except Parkal are at present irrigated by canal waters. Parkal has not yet been provided by canal water, but has vast areas under tank irrigation. This factor explains the inclusion of Parkal in high agricultural efficiency areas alongside with the canal irrigated areas. Our earlier assumption that canal irrigation becaue of its greater degree of reliability increases agricultural efficiency is thus proved correct. The high agricultural efficiency taluks are also characterised by great proportion of area sown more than once a year and lesser proportion of area under the category of culturable waste. It is the area where agricultural innovations are adopted.

Medium Agricultural Efficiency Region

This region comprises of seven taluks namely Manthani, Gangadhara, Karimnagar, Huzurabad, Chityal, Warangal rural and Nirmal. All these taluks have not yet been provided with canal irrigation facilities. However they have to some extent, efficient tank irrigation system which caters to the irrigation needs at least for one season during a year. Yields of crops are high in isolated pockets as the farmers try to apply greater amounts of inputs in the existing irrigated area. However, no effort is made to convert culturable waste lands. Proportion to total land is relatively higher than in the high agricultural efficiency region. This region is in a transition period. Once the irrigation system stabilises with the provision of canal waters, the region is bound to catch up with the High Agricultural Efficiency Region.

Low Agricultural Efficiency Region

Only two taluks namely Peddapalli and Bhimadevarapalli constitute the Low Agricultural Efficiency Region. This region does not have any appreciable proportion of area irrigated at present. Yields of crops are low and so also the area under high yielding varieties of crops. Innovative agricultural infrastructure facilities are absent and minimal. Land tenure system existing in this region is also not favourable for promoting agricultural efficiency.

REFERENCES

Bhatia, S.S. (1965): Patterns of Crop Concentration and Diversification in India, *Economic Geography*, Vol. 41.

Buck, J.L. (1967): *Land Utilization in China*, Vol. I. University of Naking.

Clark, Colin and Haswell, Margaret (1967): *The Economics of Subsistence Agriculture*, Macmillon, London.

Deshpande & Sapre (1964) : Inter-district variations in Agricultural Efficiency in Maharashtra State, *Indian Journal of Agricultural Economics*, Vol. 19, No.1.

Kendall, M.G. (1939) : The Geographical Distribution of Crop Productivity in England, *Journal of the Royal Society*, Vol. 102.

Khusro, A.M. (1964) : Returns to sale in India Agriculture, *Indian Journal of Agricultural Economics*, Vol. 19

Mohammad, Ali (1978) : *Studies in Agricultural Geography*, Rajesh Publication, New Delhi.

Shafi, M. (1960) : Measurement of Agricultural Efficiency in U.P., *Economic Geography*, Vol. 36.

Singh, Jasbir (1976): *An Agricultural Geography of Haryana*, Vishal Publications, Kurukshetra.

Stamp, L.D. (1960): *Our Developing World*, London.

Vries, E. de. (1967): Referred by Ali Mohammad (1978, pp. 74), *Studies in Agricultural Geography*, Rajesh Publications, New Delhi.

FOUR

Settlements : Distributional Pattern and Functional Characteristics

A peep into the history of mankind shows man's constant struggle against the problems posed by the nature. The involvement of man with nature manifests in many forms. Settlements are one such form. Human settlements owe their origin to the time when the human race shed its nomadic culture and established sedentary cultivation. As man made further progress in economic, social and cultural spheres, settlements began to acquire new traits, reflective of the transforming society and economy.

Conceptually, settlements are explained as places of human habitation and resultant features of man's interaction and association with nature. Structurally, settlements derive certain characteristics like, distribution, size, growth and functional activities. These attributes of settlements express existing reality of the economic order, direction and framework. The settlement system is a historic product of social, economic, political, cultural, locational and physical, realities and exhibit uneven characteristics. Consequently, settlements spread on the global plane, range from small isolated dwellings to more advanced and complex city structures.

An examination of settlements of Sriramsagar Command Area unravels complex and vibrant processes. Prior to the year 1970, the type of socio-economic evolution of settlements was similar to that of any other backward regions of Telangana and the country at large. Gradual and passive evolutionary process is replaced by more dynamic and revolutionary processes after the provision of assured irrigation facilities. Economy has become vibrant and exerting its influence in many spheres of human activities. A study into the settlement structure is therefore, essential in the light of rapidly transforming economy.

Size-Distribution of Settlements

The Sriramsagar Command Area has a total of seven hundred and forty-three inhabited rural settlements. Besides, there are twelve urban settlements and twenty-two uninhabited settlements (1981). Spatial aspects of settlement distribution is examined here with population size as the criteria. Analysis is made of the relative strength of different sized settlements on taluk basis. The size classes considered are:

Class	*Population Range*
I	Above 10,000
II	5001 - 10,000
III	2001 - 5,000
IV	1001 - 2,000
V	1000 and less

The average size of a settlement in Sriramsagar Command Area is 3499 which is more than double the average size of an Andhra Pradesh village (1523). There has been not much a discernible increase in the average size during the course of last ten years. Average size of a village in Sriramsagar Command Area increased by 4.8 per cent during the period 1971-81, whereas, the same grew by 17.0 per cent in A.P. The density pattern of settlements in the command area differs markedly with that of the State. Sriramsagar Command Area had 9.8 settlements per hundred square kilometers of area in 1981, as against 4.2 settlements for the same area in Andhra Pradesh. This is so, because the average size and density of settlements are inter-linked. Lack of substantial strengthening of settlement size in Sriramsagar Command Area may be due to the non-percolation of all the benefits of irrigation and other factors

A perusal of Table 4.1 indicates an overwhelming number of settlements have population of less than five thousand people. In quantitative terms, 95 per cent and 96.5 per cent of settlements had population of less than five thousands in 1981 and 1971 respectively.

However, some noteworthy trends emerge from a cursory examination of Table 4.2. It can be noted that the share of small villages with less than thousand population decreased markedly during the last censal decade, whereas, the size class of 1001-2000 maintained its

TABLE 4.1 : Sriramsagar Command Area: Settlements By Size Class - 1971-81

Sl. No.	Taluk	Settlements having Population of															
		<1000		1000-2000		2000-5000		5000-10000		10000-20000		Total Rural settlements		Uninhabited settlements		Urban settlements	
		1971	1981	1971	1981	1971	1981	1971	1981	1971	1981	1971	1981	1971	1981	1971	1981
1.	Karimnagar	13	8	12	12	13	15	–	2	–	2	38	39	–	1	1	1
2.	Metpalli	8	4	21	13	10	20	4	3	2	2	45	42	2	5	2	2
3.	Jagtial	39	34	50	45	27	37	4	4	–	1	120	121	1	–	1	1
4.	Peddapalli	56	48	72	68	50	56	5	11	–	–	183	183	5	5	3	3
5.	Huzurabad	16	10	27	25	36	43	5	5	–	1	84	84	1	–	–	1
6.	Manthani	28	26	10	10	4	6	–	–	–	–	42	42	2	2	1	1
7.	Parkala	14	11	19	20	18	19	1	2	–	–	52	52	–	–	–	1
8.	Warangal	10	8	9	11	33	33	3	3	–	–	33	33	–	1	–	–
9.	Lakshettipet	27	25	22	19	8	14	–	–	–	57	58	4	1	–	–	–
10.	Nirmal	22	17	6	7	2	6	–	–	–	–	30	30	4	4	1	1
11.	Khanapur	29	23	11	14	5	7	1	2	–	–	46	46	2	–	–	–
12.	Armur	3	–	8	8	2	4	–	–	–	–	13	13	–	–	1	1
	SRCA	265	217	267	252	186	238	23	32	2	4	743	743	21	22	11	12

TABLE 4.2 : **Sriramsagar Command Area : Proportion of Settlements : A Size Class Analysis - 1971-81**

Sl. No.	Taluka of 1971†	Percentage of Settlements to Total Settlement by									
		< 1000		1000-2000		2000-5000		5000-10000		10000-20000	
		1971	1981	1971	1981	1971	1981	1971	1981	1971	1981
1.	Karimnagar	34	21	32	31	34	38	-	5	-	5
2.	Metpalli	18	9	47	31	22	48	9	7	4	5
3.	Jagtial	33	28	42	37	22	31	3	3	-	1
4.	Peddapalli	31	26	39	37	27	31	3	6	-	-
5.	Huzurabad	19	12	32	30	43	51	6	6	-	1
6.	Manthani	67	62	24	24	9	14	-	-	-	-
7.	Parkala	27	21	37	39	35	36	1	4	-	-
8.	Warangal	30	24	27	33	33	33	10	10	-	-
9.	Lakshettipet	47	43	39	33	14	24	-	-	-	-
10.	Nirmal	73	57	20	23	7	20	-	-	-	-
11.	Khanapur	63	50	24	30	11	15	2	5	-	-
12.	Armur	23	8	62	61	15	31	-	-	-	-
	SRCA	35.67	29.0	35.93	34.0	25.0	32.0	3.1	4.0	0.27	1.0

share to the total Sriramsagar command area. The proportion of villages within the population size class of 2001-5000 registered significant upward movement. It can, therefore, be safely deduced that the large villages grew at the cost of small villages. The marginal fall in the number of settlements of size class 1001-2000 and conspicuous fall in the settlements of the size smaller to this class, though universal in the proportion of settlements with a population of less than two thousands and very significant increase in the size class of 2001-5000 can be noticed in the irrigated component of the command area. The largest size class of settlements with above 10,000 population in command area was conspicuously absent in 1971 in all the taluks except in the lone taluk of Metpalli. But in 1981, the picture was altered with the entry of villages belonging to this size class in some taluks like Karimnagar, Jagtial and Huzurabad, besides Metpalli.

Nearest Neighbour Analysis of Settlements Distribution

Settlements distribution on the space varies from one place to the other. With a naked eye, if one observes the distribution phenomena, one calls it either dense or sparse which statement is subjective, qualitative and unscientific. To overcome this methodological fallacy, mathematical treatment is desirable. One such technique which can be used to determine the spatial randomness is Nearest Neighbour Index. The index explains the settlement distribution quantitatively and provides a continuous scale starting with '0' and ending with 2.15. With this index, any distribution can be assigned a measure and it can be fitted in the distribution pattern. As this method is developed on mathematical principle, it is immune from illogical fallacy of subjectivity. The index is referred here as 'R'. The distribution of settlement can be explained by three concepts. If the index value is '0', the distribution is complete clustering; likewise suppose it comes 'I', the distribution is said to be random; and if it results with an index value of 2.15 the distribution is called 'even and regular'.

An examination of the Table 4.3 reveals that the 'R' value is 1.22 when all the settlements are considered as a whole in the command area irrespective of size classes, it is close to random distribution. Intra-taluk

TABLE 4.3 : **Settlements : Spatial Distribution Pattern**

Sl. No.	Name of the Talukas	Nearest Neighbour Index for Settlement with Population					
		Total Settlements	<1000	1000-2000	2000-5000	5000-10000	Urban Setl.
1	2	3	4	5	6	7	8
1.	Armur	1.24	–	1.88	1.28	–	–
2.	Bhirngal	1.40	–	1.33	0.93	–	–
3.	Metpalli	1.53	1.24	1.24	1.31	1.69	–
4.	Jagtial	0.84	0.73	0.81	0.76	1.02	–
5.	Mallial	1.36	1.26	1.28	1.26	–	–
6.	Seektanabad	1.41	0.43	1.17	1.42	1.32	–
7.	Peddapalli	1.15	1.08	0.52	1.06	1.24	–
8.	Manthani	1.16	1.25	1.38	1.41	–	–
9.	Gangadhara	1.63	–	1.52	2.13	–	–
10.	Karimnagar	1.46	1.54	1.05	1.20	–	–
11.	B.D'palli	1.40	1.01	0.75	1.41	–	–

12.	Huzurabad	1.54	1.40	1.15	1.30	1.53	-
13.	Chityal	1.83	-	1.33	1.47	-	-
14.	Parkal	1.53	1.94	1.45	1.53	0.95	-
15.	Warangal	1.52	1.56	0.71	1.20	-	-
16.	Nirmal	1.19	1.51	0.98	1.19	-	-
17.	Khanapur	1.32	1.28	1.34	1.42	0.20	-
18.	Lakshettipet	1.06	0.96	0.86	1.12	-	-
	Command Area	1.22	0.95	0.89	1.13	1.13	1.50

R = 0 complete clustering = 1 Random distribution = 2.1491 Even distribution analysis reveals that 'R' value ranges from 0.84 to 1.83. Although, some values are close to random distribution, many of them are placed almost in the mid-way towards even distribution. Only one taluk namely, Chityal appears to be nearer (1.83) to even distribution.

An analysis of the settlements with less than thousand population explains that, by and large the settlements demonstrate random distribution. The index for command area for the size class is 0.89 which is in no way near to the distribution pattern of all settlements considered together. Parkal taluk being 1.94 is the extreme and Jagtial has the lowest 'R' value of 0.73. All the other values range inbetween, Settlements having population size of 1001 to 2000, too, demonstrate almost similar characteristics. Nearest neighbour Statistic for the command area of this class is 0.89 which is again falling close to random distribution. Specially, in the taluks of Jagtial, Bhimadevarapalli, Warangal, Nirmal and Lakshettipet; it is almost just above that of command area. All others have values ranging between random to even distribution.

The index value is 1.13 for the settlements with population size class of 2001 to 5000 and 5001 to 10000 in the command area. The taluks like Metpalli and Huzurabad have nearest Neighbour Index of 1.69 and 1.53 respectively, and Sultanabad and Peddapalli have similar measures of 1.24 and 1.32. In the size class of 2001 to 5000, many taluks, Armur, Metpalli, Sultanabad, Manthani, Bhimadevarapalli, Huzurabad, Chityal Parkal, Khanapur are placed in the index range of 1.28 to 1.53. In contrast to all others, Gangadhara demonstrated almost the character of even distribution with a measure of 2.13.

It is evident from an analysis of Nearest Neighbour statistics in urban settlements that the 'R' value is 1.5 which is almost inbetween random and even distribution. An observation of all settlements explains that in many taluks the 'R' value is fluctuating between 1 and 2.15, hence it can be inferred that settlements are dispersed from random towards even distribution which results in a sort of regularity in its spatial distribution pattern of settlements.

Growth Pattern of Settlements

Growth is an inherent character of human society. As

the settlements stand for inhabitation, they do reflect the changing nature of population. A physical change is manifested in the settlement sizes as a sequel to this dynamic phenomenon. A close examination of Table 4.4 reveals that a constant shift is taking place in settlement size classes. The phenomenon can be seen by observing the movement of lower size class settlements into the succeeding group. The process is continuous and has a chain effect.

Growth of the settlements is worked out by adopting the following formula:

$$r = \frac{P_n - P_o}{P_o \times n} \times 100$$

where

r = annual amount of change or rate of change

P_o = initial population

P_n = Population at the end of the period

n = time in years.

The formula has been adjusted to the decadal growth rate by multiplying the years of a decade and the data are interpreted. The analysis of the settlement size classes of command area during 1971 to 1981 reveals that a clear picture is discernible in the growth structure. Almost all the settlements with population size of less than 1000 and more than half of the settlements of succeeding class clearly showed a negative growth rate. Consequently, the Smaller Settlements moved into the succeeding class after attaining the population increase. The increase is moderate with regard to the medium order settlements. The bulging is very rapid in the largest sized settlements, and some of these are attaining urban characteristics.

Intra-taluk study denotes that the decrease in the lowest order settlements range from -7 per cent in every backward taluk to -50 per cent in fast growing in Metpalli. On an average, the command area reflects -18.1 per cent decline. A fast erosion is also found in Karimnagar, Huzurabad and Armur taluks. Moderate decline is evident in Warangal, Nirmal, Parkal, and Khanapur taluks. Relatively less depletion is taking place in Jagtial,

TABLE 4.4 : **Sriramsagar Command Area: Settlements Decadal Growth Rate, 1971-81**

Sl. No.	Taluks of 1971	Percentage of Settlements to Total Settlements by				
		<1000	1000-2000	2000-5000	5000-10000	10000-20000
		1971-81	1971-81	1971-81	1971-81	1971-81
1.	Karimnagar	38.5	0	15.4	200.00	200.00
2.	Metpalli	50.0	38.0	100.00	33.3	0
3.	Jagatial	12.8	10.0	37.0	0	100.0
4.	Peddapalli	14.3	5.5	12.0	120.0	0

Peddapalli and Lakshettipet taluks. On an average, the settlements with population size of 1001 to 2000 demonstrate a negative growth of -6 per cent. The trend is more conspicuous in Metpalli, Peddapalli, Jagtial and Huzurabad taluks. A slight positive change in apparent in Parkal, Warangal, Lakshettipet, Nirmal and Khanapur taluks but no alteration is recorded in Karimnagar, Manthani and Armur taluks.

Interestingly, in the size class of 2000 to 5000 population, the trend is reversed and an increase in growth is glaring and this holds good in the settlement class of 10,001 to 20,000 also. A very fast growth is displayed by Metpalli, Nirmal and Armur taluks, ranging from 100 to 200 per cent in the population size class of 2000 to 5000. A relatively moderate growth is visible in all other taluks except in Warangal, where, no sign of change is marked.

In the settlement size group of 5000 to 10,000 the growth rate exhibited by command area is 39 per cent. Of all the Command Area Karimnagar Peddapalli, Parkal and Khanapur are the fastest growing taluks, whereas in Jagtial, Huzurabad and Warangal the change is non-existent. Manthani, Lakshettipet, Nirmal and Armur do not figure in the size class. The growth rate of the size class of 10,000 to 20,000 of Karimnagar, Jagtial and Huzurabad is very high and almost considered with overall performance of command area. The rest do not get projected in this size class.

Having scanned through the different size classes of the settlements, it can be inferred that the region is passing through a process of rapid transformation. The study establishes two contradictory trends of having negative growth at one end and a very fast growth at the other end of the spectrum of population size classes of Sriramsagar Command Area.

Distributional Pattern of Settlement Functions

In a developing economy, rural functions play a vital role. Since the inception of assured irrigation, the command area is emerging as the dominant agricultural region. A strong infrastructural base is necessary in the light of changing socio-economic structure. As the infrastructure and agricultural system in Sriramsagar Command Area are compatible to one another there is a need for a close examination of the distributional pattern of infrastructural facilities, in the context of integrated development of the command area.

The fundamental functional pre-requisites are : medical and health, education, agricultural services, transport, road network, postal communication banking, marketing, veterinary services etc.

Education

Education is one of the few vital instruments which help in accelerating the socio-economic transformation. The population needs to be properly educated, if, the fruits of economic development are to reach every one in an equitable manner. Mere construction of dams and provision of assured irrigation facilities would not automatically lead to reaping of optimal social and economic harvests from such huge investments. It is, therefore, imperative on the part of authorities to raise the level of literacy and educational status of the population. If the people are properly educated and made aware of the recent technological innovations in agricultural sector, the productivity levels in the command area are bound to register an upward movement. Education is required not only from an economic point of view, but also for making people socially conscious and aware of their rights. Education of the masses would normally lead, atleast in the long run, for a more just and equitable distribution of nations wealth.

Sriramsagar Command Area is moderately endowed with educational facilities of different levels. Primary school, being the lowest service in the educational hierarchy, is found in almost all settlements excepting in a few villages. In places where primary school is not found, it is either substituted by Upper Primary and High School where Primary school sections are also included.

Although the Upper Primary School, which occupies second rung from below in the education ladder, do not exhibit any central character in the Indian scene as it is ubiquitous in spatial distribution. An observation of Table 4.5 shows that its distribution is limited to certain settlements of the command area. More concentration of Upper Primary Schools range from 50 to 70 per cent of settlements is noticed in and around Karimnagar.

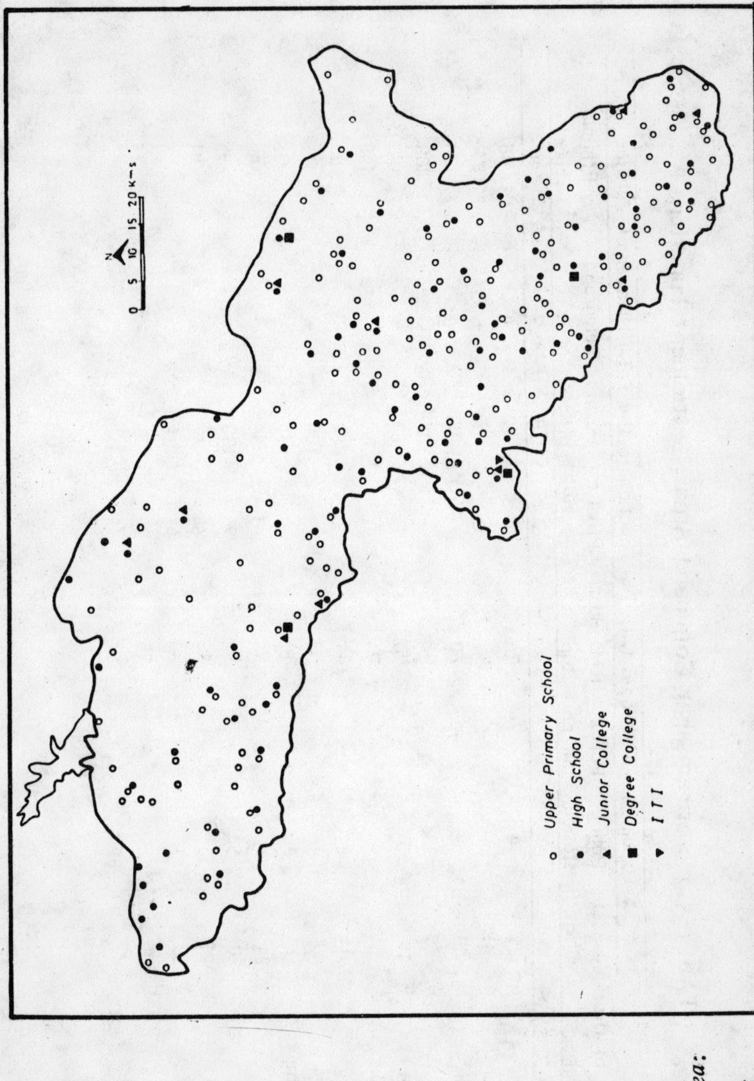

FIG. 4.1: *Sriramsagar Command Area: Educational Facilities 1985*

TABLE 4.5 : **Sriramsagar Command Area : Settlement Functions**

Sl. No.	Name of the Mandal	% Population of Villages with the Function to the Total Number of Villages							
		Mandal village	Police station	Police out post	RTC Bus stop	Private bus stop	Post office	Telegraph office	Primary school
1	2	3	4	5	6	7	8	9	10
1.	Karimnagar	25	4	-	76	4	72	4	88
2.	Manakondur	18	-	-	94	-	94	-	83
3.	Chopadandi	10	-	-	60	-	80	-	100
4.	Huzurabad	9	11	-	18	-	100	11	100
5.	Jammlkunta	19	5	-	52	-	89	5	100
6.	Venenavanka	14	-	-	71	-	86	-	100
7.	Kamalapur	16	-	-	44	-	100	-	100
8.	Elkathuthi	9	-	-	33	-	100	-	100
9.	Keshavapatnam	17	6	-	82	-	76	-	76
10.	Jagtial	24	4	4	54	63	96	4	100
11.	Raikal	20	5	-	75	90	100	5	95
12.	Dharmapuri	22	5	-	55	64	79	5	81
13.	Saranapur	23	-	-	61	61	78	-	74
14.	Mallial	4	-	-	75	25	75	-	75

15. Pegadapalli	12	–	–	67	58	100	–	100
16. Gollapalli	22	5	–	72	64	95	–	100
17. Metpalli	4	25	–	75	75	75	25	100
18. Koratla	15	7	–	47	100	87	7	100
19. Medipalli	6	–	–	67	17	83	–	100
20. Ibrahimpatnam	14	–	–	64	29	57	–	79
21. Mallapurr	22	–	–	77	82	64	–	59
22. Peddapalli	23	4	–	48	13	74	4	96
23. Srirampur	17	–	41	41	88	6	100	35
24. Odela	11	9	–	45	45	90	9	100
25. Ramagundam	30	3	–	27	23	73	3	100
26. Kamanpur	25	4	–	28	24	48	4	88
27. Sultanabad	21	5	–	–	71	33	10	5
28. Velgatur	24	–	–	46	42	71	–	83
29. Dharmaram	17	6		41	65	76	–	88
30. Julapalli	15	–	–	53	73	93	–	80
31. Manthani	35		–	31	9	37	3	86
32. Mutharam	18	–	–	44	–	56	–	94
33. Nirmal	5	20	–	80	–	80	20	60

(Contd.)

TABLE 4.5 : *Contd.*

1	2	3	4	5	6	7	8	9	10
34.	Laxmanchanda	13	-	-	77	-	62	8	69
35.	Mamda	10	-	-	70	-	70	-	60
36.	Khanapur	17	6	-	53	-	41	6	53
37.	Kaddam	24	-	4	92	-	33	4	92
38.	Jannaram	21	5	5	43	-	52	5	90
39.	Dandapalli	27	-	-	46	-	46	-	100
40.	Lakshettipet	12	-	-	50	-	42	-	92
41.	Chityal	3	-	-	-	33	33	33	67
42.	Mogulapalli	19	-	5	-	11	79	-	74
43.	Parkal	23	4	-	70	30	78	4	65
44.	Shyampet	6	17	-	50	-	83	-	50
45.	Atmakur	15	7	-	73	47	80	-	93
46.	Hasamparthi	15	7	-	87	27	93	-	100

Sl. No.	Name of the Mandal	% Population of Villages with the Function to the total number of villages									
		Upper Primary school	High School	Junior College	Degree College	Primary health centre	Hospital	Rural Health centre Allopathic	Primary Health centre Others	Private Doctors Allopathic	Market yard
1	2	11	12	13	14	15	16	17	18	19	20
1.	Karimnagar	52	28	4	4	4	4	8	4	12	12
2.	Manakondur	44	28	-	-	6	-	6	6	22	11
3.	Chopadandi	70	30	10	-	-	-	-	10	80	10
4.	Huzurabad	78	22	11	-	-	11	-	-	22	11
5.	Jammikunta	58	26	-	5	-	-	5	5	-	5
6.	Venenavanka	64	14	-	-	-	-	7	7	36	-
7.	Kamalapur	63	31	-	-	-	-	6	-	44	6
8.	Elkathuthi	44	11	-	-	-	-	-	11	33	-
9.	Keshavapatnam	47	18	-	-	-	-	-	6	-	-
10.	Jagtial	-	4	4	4	-	4	4	4	4	4
11.	Raikal	20	10		-	5	5	-	-	5	10
12.	Dharmapuri	9	9	9		-	-	5	-	14	5
13.	Saranapur	13	-	-	-	-	-	-	-	-	-
14.	Mallial	75	-	-	-	-	-	-	-	25	-

(Contd.)

TABLE 4.5 : *Contd.*

1	2	11	12	13	14	15	16	17	18	19	20
15.	Pegadapalli	42	17	-	-	4	-	8	17	42	17
16.	Gollapalli	14	5	-	-	-	-	-	-		5
17.	Metpalli	50	25	25	-	-	25	-	5	25	25
18.	Koratla	40	27	7	7	-	7	-	47	53	7
19.	Medipalli	33	33	-	-	-	-	-	50	50	17
20.	Ibrahimpatnam	14	7	-	-	-	-	-	36	29	-
21.	Mallapurr	23	5	-	-	-	-	-	-	41	-
22.	Peddapalli	48	35	4	-	-	4	4	43	17	4
23.	Srirampur	12	-	-	-	-	-	-	35	12	-
24.	Odela	45	36	-	-	-	-	18	45	45	9
25.	Ramagundam	17	3	3	3	-	7	7	20	7	7
26.	Kamanpur	32	8	-	-	-	-	-	16	12	8
27.	Sultanabad	-	-				10	-	24	24	10
28.	Velgatur	21	4	-	-	-		-	-	33	-
29.	Dharmaram	18	12	-	-	-	-	-	-	53	6
30.	Julapalli	47	27	-	-	-		7		67	7
31.	Manthani	14	6	3	3	3	3	3	23	6	13
32.	Mutharam	28	-	-	-	6	-	-	22	-	-
33.	Nirmal	40	20	-	-	-	-	-	-	-	-

34. Laxmanchanda	-	15	-	-	8	-	-	8	-	-
35. Mamda	10	20		-	-	-	-	-	-	-
36. Khanapur	35	6	6	-	-	6	-	6	-	-
37. Kaddam	8	8	-	-	-	-	-	-	-	4
38. Jannaram	10	10	-	-	5	-	-	10	-	-
39. Dandapalli	17	-	-	-	-	-	-	-	-	8
40. Lakshettipet	17	8	-	-	-	-	-	-	-	-
41. Chityal	-	-	-	-	-	-	-	-	-	-
42. Mogulapalli	26	11	-	-	-	-	11	-	-	5
43. Parkal	22	13	4	-	-	4	-	-	4	4
44. Shyampet	50	17	-	-	17	-	-	-	-	-
45. Atmakur	40	13	-	-	7	-	-	-	-	7
46. Hasamparthi	53	13	-	-	-	-	-		73	-

(Contd.)

TABLE 4.5 : *Contd.*

Sl. No.	Name of the Mondal	% Population of Village with the Function to the Total Number of Villages										
		Market	Commercial Bank	Co-operative bank	Agri. chem. & fert. Dept.	Agricultural service centre	Veterinary centre	Agricultural co-op. society	Fair price shop	Electrified villages	Pucca Road	Rest House
1	2	21	22	23	24	25	26	27	28	29	30	31
1.	Karimnagar	4	8	4	8	4	32	16	92	92	84	4
2.	Manakondur	-	17	-	22	-	39	33	94	100	94	-
3.	Chopadandi	10	20	10	-	-	30	100	100	00	60	-
4.	Huzurabad	11	22	11	11	-	44	56	100	100	44	11
5.	Jammikunta	5	11	5	5	-	43	64	100	100	37	5
6.	Venenavanka	-	14	-	-	-	21	14	100	100	14	-
7.	Kamalapur	-	19	6	6	-	31	44	88	100	19	6
8.	Elkathuthi	-	-	-	-	-	22	11	100	100	22	-
9.	Keshavapatnam	-	12	-	-	-	18	29	94	100	35	6
10.	Jagtial	4	8	4	4	4	23	29	100	100	100	4
11.	Raikal	5	-	5	-	-	-	5	30	100	100	5
12.	Dharmapuri	5	9	5	5	-	5	27	91	95	77	9
13.	Saranapur	-	9	-	-	-	17	26	100	96	78	4
14.	Mallial	-	-	-	-	-	-	50	50	75	75	-

15. Pegadapalli	-	8	-	-	-	-	42	92	100	58	-
16. Gollapalli	5	9	-	-	-	-	22	64	100	86	-
17. Metpalli	25	50	25	25	-	25	50	75	75	50	25
18. Koratla	7	20	7	7	-	53	53	93	100	60	7
19. Medipalli	-	17	-	-	-	17	50	83	100	50	-
20. Ibrahimpatnam	7	7	-	-	-	14	43	100	100	43	-
21. Mallapurr	-	14	-	-	-	23	32	77	91	-	45
22. Peddapalli	4		4	4	4	4	26	96	26	96	4
23. Srirampur	-	6	-	12	-	24	24	100	94	41	6
24. Odela	9	9	9	9	-	55	18	90	45	100	-
25. Ramagundam	3	10	-	3	-	13	17	83	96	26	3
26. Kamanpur	-	8	-	4	-	8	8	52	88	28	-
27. Sultanabad	5	10	5	5	-	19	24	81	100	62	5
28. Velgatur	-	-	-	-	-	17	17	71	92	63	4
29. Dharmaram	6	12	-	6	-	29	18	82	100	88	-
30. Julapalli	-	20	-	-	-	13	20	93	100	80	-
31. Manthani	3	6	3	3	-	9	9	46	94	11	6
32. Mutharam	-	-	-	-	-	11	17	17	94	17	-
33. Nirmal	-	40	-	-	-	20	40	80	100	80	-

(Contd.)

TABLE 4.5 : *Concld.*

1	2	21	22	23	24	25	26	27	28	29	30	31
34.	Laxmanchanda	-	15	-	-	-	23	31	85	100	77	8
35.	Mamda	-	20	-	-	-	20	20	60	100	80	-
36.	Khanapur	6	6	6	6	-	12	24	76	65	47	6
37.	Kaddam	-	8	-	-	-	8	21	63	88	83	8
38.	Jannaram	-	5	-	-	-	19	19	76	86	52	10
39.	Dandapalli	4	-	-	-	-	25	17	88	96	91	4
40.	Lakshettipet	-	8	-	-	-	19	42	83	100	75	-
41.	Chityal	-	-	-	-	-	-	100	00	100	-	-
42.	Mogulapalli	-	5	-	-	-	11	100	00	100	-	-
43.	Parkal	4	13	4	4	-	4	52	100	100	30	4
44.	Shyampet	17	-	-	-	-	17	42	83	83	17	-
45.	Atmakur	-	27	7	13	-	13	33	100	100	100	-
46.	Hasamparthi	-	7	-	-	-	27	47	100	100	100	-

Metpalli, Peddapalli, Dharamaram, Keshavapatnam, Odela and Hasanparthy 45 to 50 per cent of the settlements have upper Primary schools. In the rest of the Mandals, less than 40 per cent settlements have upper primary schools. In a great majority of mandals, upper primary schools are found in 9 to 30 per cent of the villages. The generation of additional economy in the region is not adequately supported by infrastructure like, education and this is more prevalent in areas which come under only assured irrigation system. A survey conducted by the research team reveals that a few people who are socially advanced do not want services to come into the villages as this would effect labour availability and prevent exploitation by resistance in asserting their democratic rights.

In a few mandals, the high school settlements ratio is around one-fourth to one-third of the settlements. These settlements are found in the areas with district headquarters and few former taluk headquarters. The ratio ranges between 4 to 13 per cent in a number of mandals in the command area.

In the given socio-economic framework junior college is considered a central function. In view of the constraints, it is necessary that Junior College should be located in all the Mandal Headquarters besides certain other prominent settlements. A glaring lapse is that almost two-thirds of the mandals do not have any Junior College function. The ones which have the former headquarters and a few other industrial towns. Immediate steps are necessary to augment the facilities in order to fulfil the present and future needs of the Command Area.

Degree being a central and specialised function need to be represented evenly. Only few settlements have this function and few more can be added to make things easier and to meet the present and future demands.

Medical and Health Services

Medical and Health Service is a fundamental pre-requisite of any human civilisation. Its importance is being felt as the society advances. It is necessary to examine closely the functional distribution as the study area is passing through a radical transformation. Medical and Health as a service is explained in an unified form in which we find various levels of functions. As the importance and inputs of the units increase, the distribution is restricted to a few in number by serving more tributary area. The primary function of medical and health rests with a private doctor and proceeds in the form of organised units of various orders starting with Rural Health Centre. Primary Health Centre to the higher order function like full-fledged hospital. In the light of multiplicity in structure, the data pertaining to this service are processed by taking individual functional units (Fig. 4.2).

Private allopathic doctors function being in unorganised sector and left to the descretion of individuals it is they who determine the location of the function. It is nothing to do with policy functions which are the brain-child of government sector. Although, the function is available in all mandals, it is not found in all the settlements. Its density is more in two prominent areas; one being an area which came under irrigation during 1970s and the other is in areas located close to the important urban centres. The common criterion is buoyancy and vitality of the economy.

Establishment of rural health centre is a step in the right direction in rural India. A very low number of settlements have the facility, particularly, the allopathic one. All the mandals have this facility, with some exceptions. These functions can be popularised and made available for many more settlements. There is a good amount of area served inadequately and there are mandals which are neither served by allopathic units nor others. Facilities should be provided on a priority basis to such areas.

Primary health centre is a function in the country-side which is equipped with a qualified doctor supported by modern inputs. Proper diagnostic facilities are made available to village masses and it is the only available service in relatively inaccessible areas. Table 4.6 reveals that this vital function is mostly ignored as is not found in many parts of mandals. However, it is proposed by the Government that each and every mandal should be equipped with primary health centre by adding additional facilities to the existing ones.

Hospital appears to be mostly confined to the urban areas of the command area. As the hospitals provide better opportunity for people to get treated by the government should have greater stress to take them into the

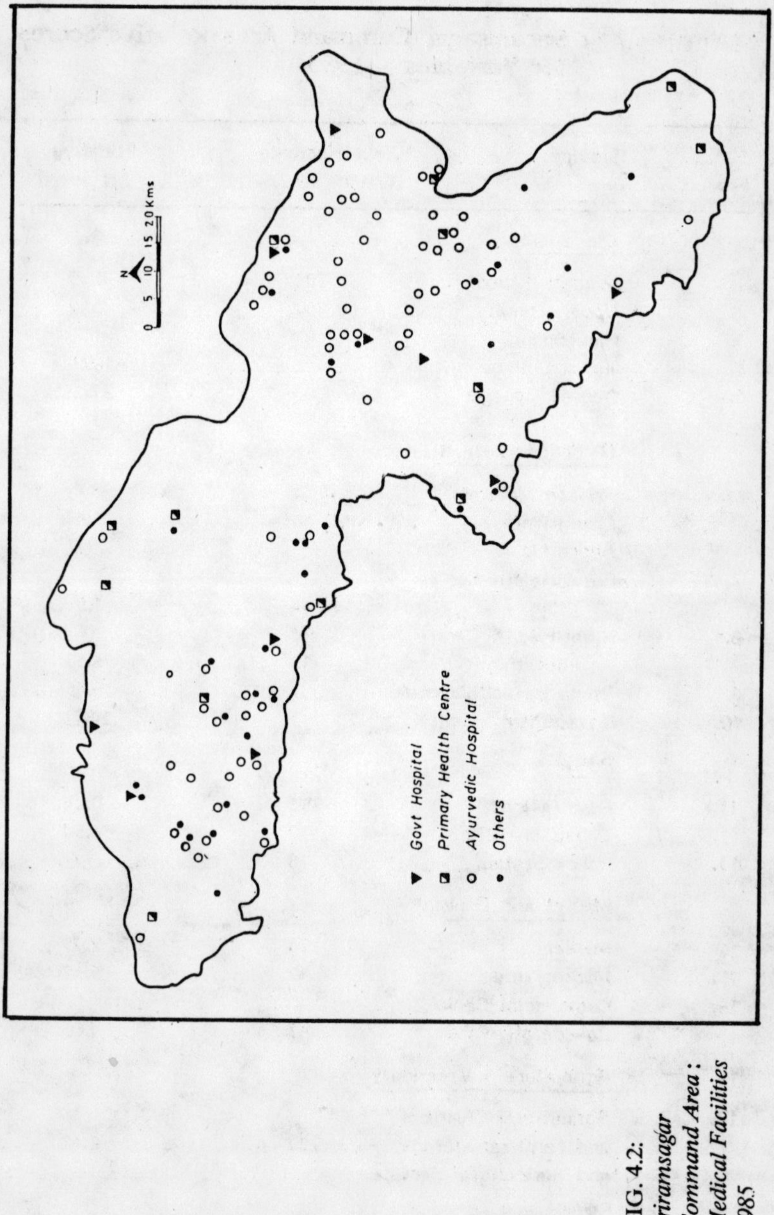

FIG. 4.2: *Sriramsagar Command Area : Medical Facilities 1985*

TABLE 4.6 : **Sriramsagar Command Area: Relative Scores for Variables - 1985**

Sl. No.	Services	Settlements with service	Relative Service
	Educational		
1.	Primary School	201	0.33
2.	Upper Primary School	157	0.42
3.	High School	81	0.78
4.	Junior College	6	10.92
5.	Degree College	3	21.76
	Medical and Health		
6.	Private Medical Practitioner Allopathic	96	0.65
7.	Rural Health Centre others	44	1.5
8.	Rural Health Centre Allopathic	10	6.53
9.	Primary Health Centre	7	9.35
10.	Government Hospital	3	21.76
	Postal		
11.	Post Office	255	0.26
12.	Telegraph Office	10	6.53
13.	Police Station	18	3.66
	Market and Bank		
14	Market	24	2.75
15.	Market yard	12	5.42
16.	Commercial Bank/ Co-operative Bank	63	1.05
	Agriculture & Veterinary		
17.	Agricultural Chemical and fertilizer storage and Agricultural Service centre	27	2.42

Sl. No.	Services	Settlements with service	Relative Service
18.	Agricultural Co-operative society	123	0.52
19.	Veterinary Centre	85	0.78
	Others		
20.	Fair Price Shop	228	0.26
21.	Electricity	237	0.25
22.	Travellers Bunglow	31	2.09

rural areas.

Agricultural Service Centre and Agricultural Chemical and Fertilizer Depot

The region is a predominantly agricultural one. The nature of agriculture is quite different to that of the agriculture prior to say about twenty years. The lands which have been lying fallow all these years and the grass lands have emerged as the most productive agricultural lands. Farmers accustomed to age old practices had to adopt themselves to the changing needs of the system. As most of them are illiterates and strange to the fast changing scientific and technological inputs, they need greater attention by way of training and guidance. With this as a backdrop, we probe into the facilities provided in the agriculturally dominated command area. An analysis of the depots registered are inadequate. It is pathetic to find that certain mandals are left without this service. This poses a serious defficiency in the emerging framework (Fig. 4.3).

Veterinary Centre

Veterinary Centre is a prerequisite in any animal husbandry or agriculture dominated area. It's importance is stressed by the fact that most of the farming activity is carried out by bullocks. Besides, cattle and milk ani-

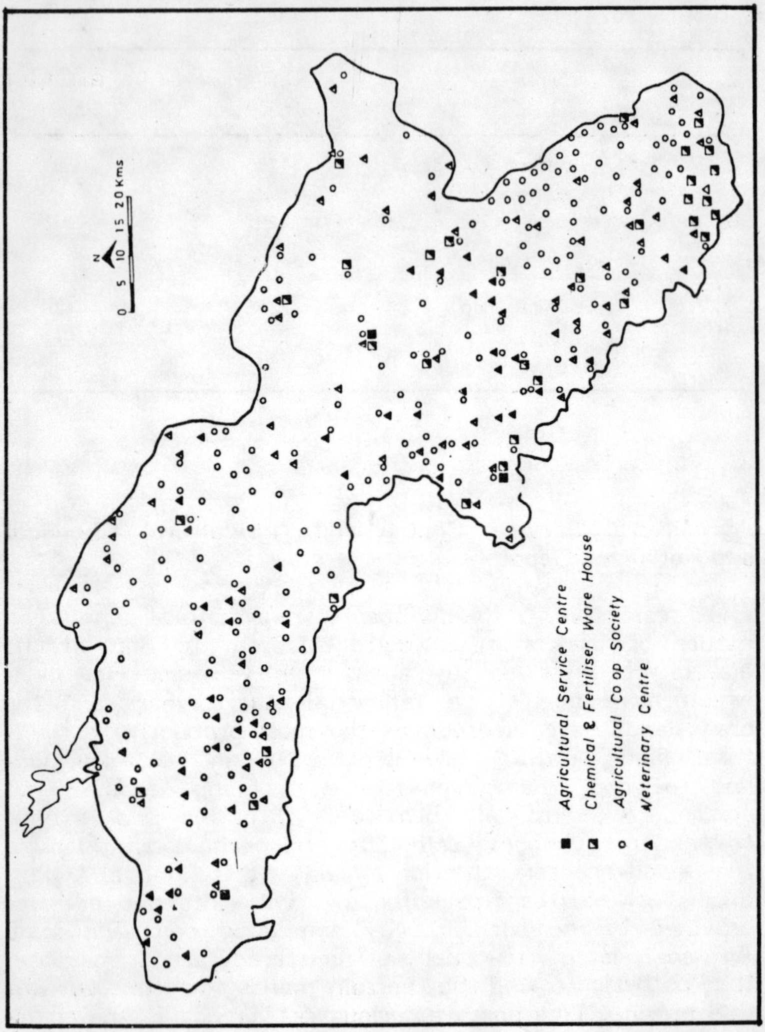

FIG. 4.3: *Sriramsagar Command Area: Agricultural Services 1985*

mals also supplement the agricultural economy. The veterinary centres are found scantily in some mandals and many others do not have the facility at all.

Market

In rural economy, market plays a major role by acting as a collecting and distributing centre. Besides, it also acts as service centre with regard to the other facilities as this acts as a transferring point of goods from by way of collecting the native produce and siphoning it into other places. It also imports goods from other areas spills them into the rural environment. A study of distribution pattern is essential in view of its greater role in the society. There are many weekly markets and market yards (regulated) in Sriramsagar Command Area. The weekly market is an off shoot of an age old marketing system. They work on particular days of the week. They still play a dominant role in rural society and economy. Almost all mandals have this facility. In the recent times, Market yard has become an instrument to lessen the irregularities, in principle at least. Their importance is being felt by producers of rural areas. The facility appears to be scanty in Sriramsagar Command Area and is required to be augmented by commissioninq a few more regulated markets. The area commissioning a few more regulated markets. The area being under rapid transformation requires greater infrastructural support. As Table 4.6 shows, there are some mandals which are neither served by weekly market nor by regulated ones (Fig. 4.4).

Rural Banking

In the changed economic environment that exists in the command area, rural banking plays vital role. It outweighs the village money-lending system by which, the farmers are relieved from the over burden of the interest rates. Banking in this area, not only serves in providing capital at cheaper interest rates to meet the agricultural inputs but also accomplishes the task of mobilising the money surpluses. This surplus can be used in rural reconstruction. The banking units found in the area are either commercial banks or co-operative banks. As for banking importance is concerned, commer-

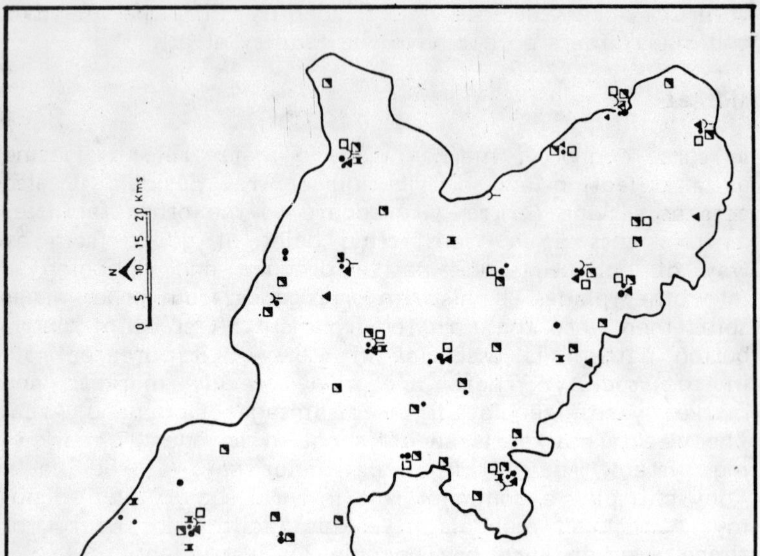

FIG. 4.4: *Sriramsagar Command Area: Market, Banking & Other Services 1985*

cial banks come into the forefront. Co-operative banks are also established in a few mandals. The picture that can be discerned from Table 4.6 is that a great many number of mandals do not have the co-operative banking system. In the emergence of a new socio-economic environment, the banking network should be strengthened and the facility can be extended to cover the weakest section of the society by adopting a judicious and efficient system of money lending.

Other Services and Facilities

Post Office facility has spread into most of the villages of the command area with 80 to 100 per cent coverage of villages. This has led to better postal communication in the area. The left out uncovered pockets should also be brought into the postal network so that no remote village is uncovered by postal facility. Telegraph office being a little higher order function, is found in a very few higher order villages of the command area. A dynamic region like SRCA requires a dense network of posts and telegraphs facilities.

In a developing socio-political environment, if the goods are not distributed properly, problems are bound to crop up creating bottlenecks in public distribution system.

Rural electrification in this area is achieved by linking almost all villages. This would help in augmenting the economy by way of its use in agriculture, industry and other activities.

Road network is the basis for any transport system. Pucca and all weather motorable roads are desirable. An examination of pucca road facility reveals that the network of roads is good in three areas: one, in northern part of Karimnagar, second, around Karimnagar and the third, in the tracts of Adilabad and Warangal districts falling in the command area. The road system is not very developed in south of Karimnagar, and backward tracts of Manthani. The availability of pucca road facility is a resultant feature of the function on command area authorities which has taken up the roads laying work. The road network is strongly related with bus services.

Man's mobility and interaction among the settlements are mostly guided by the means of transport. It is more so if the area is equipped with organised and efficient mode. The area is facilitated by the State Road Transport buses. However, we find the private network which, in some places, supplement the State, owned buses. By and large the fact remains that Road Transport Corporation is the main mode of transport in the area. Bus service links with settlements are high in places like Manakondur, Karimnagar, Keshavapatnam, Mallial, Nirmal, Parkal Mandals etc. It appears from the analysis that the old taluk headquarters and places close to district headquarters are provided the service. At one extreme, Manakondur has 94 per cent its villages covered by this facility and stands on the top, Huzurabad Mandal has just 18 per cent settlement coverage. It is all the more surprising that this area is not even provided by private bus facility. Other mandals have a service coverage of 45 to 60 per cent of settlements.

Settlement Potentialities

In the given socio-economic framework, uneven distribution of the phenomenon is an objective reality. The manifestation of dissimilarity is historically ingrained and guided by various processes. Man's influence on nature and his partisan view appropriated the functions within certain settlements units and others are left neglected. Consequently, as on off-shoot of man's thinking, emerged the potential and weak the settlements. Potentiality is studied by considering the services available in the settlements. Although services alone do not explain the total reality but it enables us to determine the relative importance as far as the service potentiality is concerned.

Methodologically, the study proceeds by taking care of subjective fallacies, method is devised here which is mathematically simple and at the same time nearer to reality. Multiple number of functions are considered by considering twenty-one variables so as to lessen the basis in the indentification of settlements potentialities. The variables taken (Table 4.6) are analysed with the hypothesis that the greater the number of functions distributed on the space, lesser is the influence on the settlements. On the other hand, if the functions are scared and the spatial spread is limited, the influence exercised by the settlement with that function will be more with the hypothesis specified, the concept of

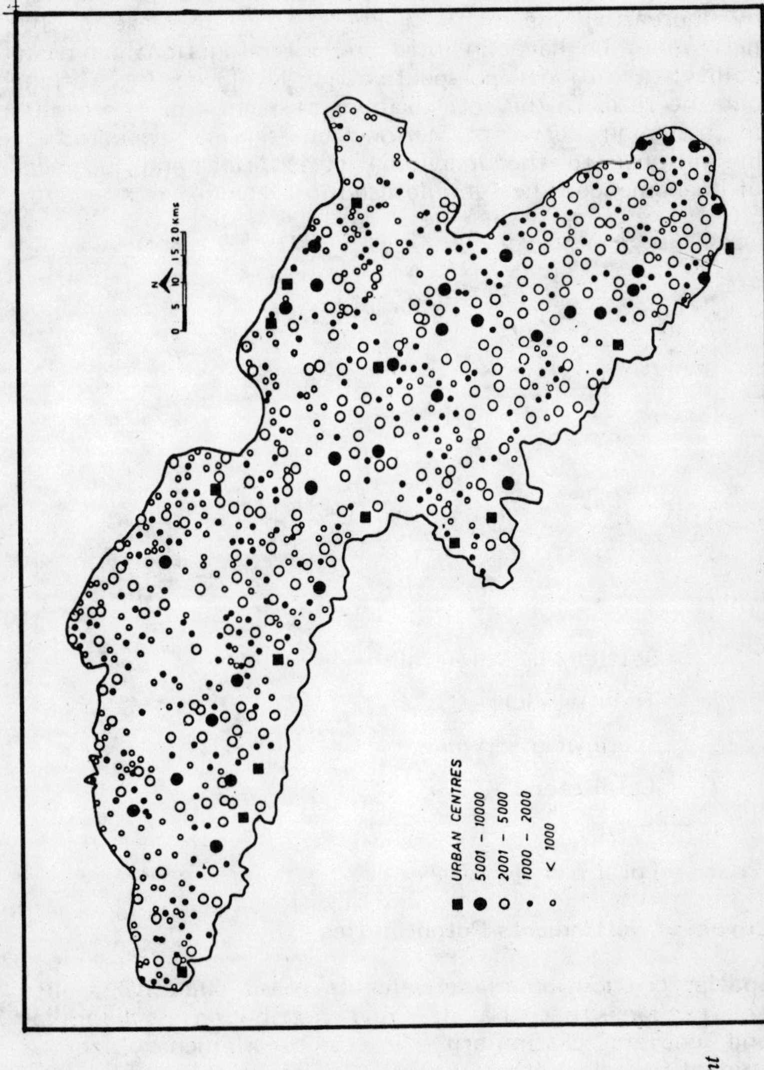

FIG. 4.5: *Sriramsagar Command Area: Levels of Settlement Population Potentiality 1981*

reciprocals is adopted as it influence the phenomenon by increasing the units if there is a decrease in the derivative. To have an intra and inter-functional comparability, a rational perspective is developed by working out the ratio to the total number of functions. The results in assigning score to a function service appeared to be rational to the individual observation and judgment of phenomenon. The formula used in technique is :

$$S_p = E \sum_{i=1}^{n} R_s \; s_i$$

$$R_s = \frac{S \times 100}{T_s}$$

$$S = \frac{1}{s_i}$$

$$T_s = E \sum_{i=1}^{n}$$

Where

S_p = Settlement Potentiality

R_s = Relative Score

S = Score of a service

T_s = Total score

s_i = Service

n = Total number of services

Levels of Settlements Potentialities

Spatial distribution of settlements is an objective reality and explains that the size and distribution is dissimilar and uneven. Dissimilarity in the settlements sizes is studied by taking the functional units of the settlements. This enables us to differentiate quantitatively. It is worked out of villages with population of 2000 and more. This can be justified because, in the rural environment of a developing economy, functions are not found significantly in the settlements with a population of less than 2000. However, and exception can be made, as in certain settlements, functions of inferior nature are found, in others a few important ones. A close examination of functional distribution of various population size classes of settlements of the command area reveals that the functions reflected in lower sized settlements are almost insignificant. Of the total settlements of the command area, 256 settlements figure in the population sizes of 2000 and above (excepting urban settlements).

The settlement potentialities are determined with the devised technique. The resulted potentiality scores range from an insignificant fraction to the highest score of 58.51. With this as the basis, the scores are plotted graphically in Fig. 4.6. Six levels of settlements potentialities emerge. The first level of settlement potentiality is marked at potentiality score of 36, and all those settlements got a score of 36 and above are considered as first level of settlements and like-wise the remaining are identified. The second score considered is from 23 to 36, third from 13 to 23; fourth 7 to 13; fifth 4 to 7 and the last below 4.

An analysis of Table 4.7 unfolds that only four settlements fall in the first level followed by five in the second, fourteen in the third, twenty-five in the fourth, thirty-one in the fifth and hundred and seventy-seven in the last. The settlements found in the level one are Khanapur, Sultanabad, Jammikunta and Dharmapuri. All the settlements in level I are found to be prominent in the regional context as they are also mandal headquarters. Khanapur and Sultanabad were once taluk headquarters and only these two settlements are unrepresented in the urban settlement category. Jammikunta being a big market place and centrally located in the region could also be justified to be in the first level. Dharmapuri is located at the river edge and also at a nodal point. By virtue of its nodal location besides being a transhipment point, it is acting as a local trading post. At the second level, the settlements represented are Gollapalli. The first three settlements are mandal headquarters. In the third level the settlements represented are Nanchera, Chilvakodur, Ganjapaduga, Sathambapalle, Lingapuram, Bhommareddypalle, Endapalle, Veldi, Motlapalli, and others which are also mandal headquarters are Velgatoor, Srirampuram, Kamanpur, Shyampet and Laxmanchanda.

The urban settlements of SRCA are Karimnagar,

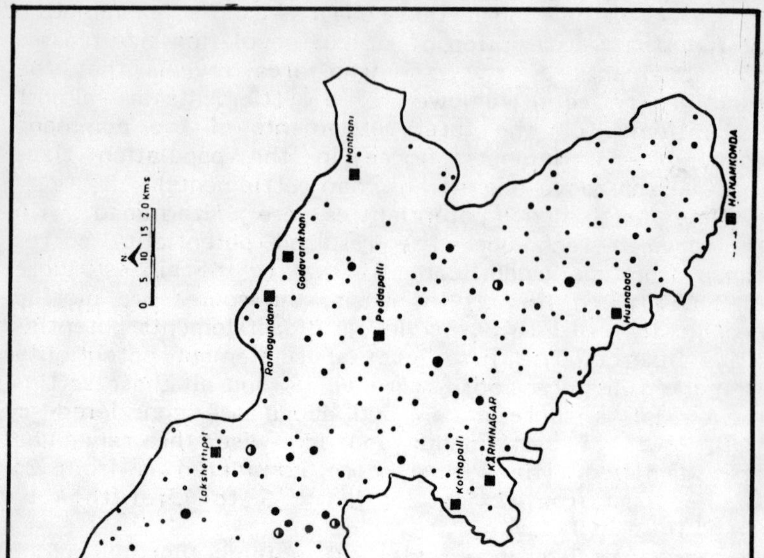

FIG. 4.6: *Sriramsagar Command Area: Levels of Service Potentiality 1985*

TABLE 4.7 : **Sriramsagar Command Area : Population Size of 2000 and Above (Excluding Urban) Settlements : Levels of Potentiality - 1985**

Sl. No.	Taluks	I	II	III	IV	V	VI	Total
1	2	3	4	5	6	7	8	9
1.	Armur	–	–	–	–	–	2	2
2.	Bhimgal	–	–	–	1	–	1	2
3.	Metpalli	–	–	–	2	6	15	23
4.	Jagtial	1	–	–	1	1	24	27
5.	Mallial	–	2	6	–	1	4	13
6.	Gangadhara	–	1	–	–	1	3	5
7.	Karimnagar	–	–	–	1	1	10	12
8.	Sultanabad	1	1	2	4	5	17	30
9.	Peddapalli	–	1	2	5	2	25	35
10.	Manthani	–	–	–	1	1	4	6
11.	Bheemadevara-palli	–	–	1	–	3	11	15

(Contd.)

TABLE 4.7 : *Contd.*

1	2	3	4	5	6	7	8	9
12.	Huzurabad	1	-	-	5	4	23	33
13.	Chityala	-	-	-	2	-	1	3
14.	Parkala	-	-	2	-	-	13	15
15.	Warrangal	-	-	-	2	2	6	10
16.	Nirmal	-	-	1	1	1	3	6
17.	Khanapur	1	-	-	-	2	4	7
18.	Lakshettipet	-	-	-	-	1	11	12
	Sriramsagar Command Area	4	5	14	25	31	177	256

TABLE 4.8 : **Sriramsagar Command Area : Settlements Potentialities - 1985**

Sl. No.	C. No.	Village Name	Population	Potential Score	Settlement level
1	2	3	4	5	6
		ARMUR			
1.	30	Mendhora	3343	2.48	VI
2.	27	Dudgaon(R)	2580	1.25	VI
		BHIMGAL			
1.	6	Yargatla	4054	11.65	IV
2.	3	Tadla Rampur	2234	2.37	IV
		METPALLI			
1.	12	Ibrahimpatnam	7052	11.45	IV
2.	61	Maidpalle (East)	3173	9.56	IV
3.	32	Paidi Madugu	3058	6.81	V
4.	40	Iyalpeer	5677	6.81	V
5.	41	Muthampeta	3002	5.52	V
6.	60	Mohonraopet	3011	5.34	V
7.	38	Chinna Metpalle	2355	4.33	V
8.	11	Godur	2683	4.13	V*
9.	16	Raghavapeta	2340	3.85	VI
10.	2	Vemullavrthi	4492	3.78	VI
11.	43	Yousufnagar	2031	3.68	VI
12.	18	Mogalpet	2776	3.68	VI
13.	42	Vempet	3751	3.1	VI
14.	39	Madhapur	2060	3.06	VI
15.	36	Katlakunta	3135	3.0	VI
16.	5	Varsakonda	3610	2.5	VI
17.	27	Chittapur	2958	1.85	VI
18.	37	Jaganpalle	2135	1.79	VI
19.	33	Mahatabpur	2421	1.63	VI
20.	34	Pormalla	3743	1.63	VI
21.	20	Damarajpalle	2693	1.63	VI
22.	44	Chawalmaddi	2278	1.11	VI
23.	23	Mallapur	6663	0.85	VI

(Contd.)

TABLE 4.8 : *Contd.*

1	2	3	4	5	6
		JAGTIAL			
1.	34	Dharmapuri	10171	43.02	I
2.	10	Dhonthaouram	3174	12.03	IV
3.	64	Allipur	5783	4.26	V
4.	31	Beerpur	2850	2.77	VI
5.	22	Vempalli-Venkatraopet	3072	2.16	VI
6.	40	Raichapalli	2814	2.08	VI
7.	24	Itkial	4198	2.05	VI
8.	11	Jaina	3256	1.72	VI
9.	21	Mottapalle		1.63	VI
10.	77	Morapalle	2163	1.63	VI
11.	83	Jabithapur	2276	1.63	VI
12.	55	Buggaram	3302	1.53	VI
13.	66	Porendla	2911	1.53	VI
14.	48	Arpapalle	2173	1.37	VI
15.	13	Thungur	3036	1.2	VI
16.	33	Rajaram	677	1.11	VI
17.	67	Polas	3548	1.11	VI
18.	76	Chelgal	3612	1.11	VI
19.	81	Motha (Rural)	2091	1.11	VI
20.	26	Ramajipal	2659	0.85	VI
21.	72	Lakshmipuram	2992	0.85	VI
22.	85	Thiramalapuram	2174	0.85	VI
23.	27	Bhoopatipuram	2343	0.78	VI
24.	78	Tatpalle		0.78	VI
25.	53	Kamalapur	2028	0.52	VI
		MALLIAL			
1.	21	Gollapalle	2971	32.03	II
2.	34	Pegadapalle	3613	25.63	III
3.	33	Nancherla	6016	22.81	III
4.	22	Chilvakadur	2549	21.58	III
5.	10	Velgatoor	2335	21.05	III
6.	28	Gunjapaduga	2193	21.05	III

1	2	3	4	5	6
7.	2	Sathambampalle	2689	20.11	III
8.	32	Tingapuram	2015	20.11	III
9.	37	Arvalla	6030	4.28	V
10.	31	Raghavapatnam	5409	1.63	VI
11.	25	Gullakota	3557	1.53	VI
12.	9	Jagdavpeta	2187	1.11	VI
13.	24	Athmakur	2003	0.85	VI
		GANGADHARA			
1.	27	Choppadandi	10072	26.95	II
2.	31	Rukmapur	2142	4.53	V
3.	7	Ragampet	5052	3.48	VI
4.	8	Arnakonda	3634	3.48	VI
5.	29	Vedhuraghattu	2787	2.7	VI
		KARIMANAGAR			
1.	30	Manakonour	7691	11.98	IV
2.	28	Gangapalli	4154	6.26	V
3.	6	Nagnur	3640	3.61	VI
4.	12	Moqdoompur	2524	3.09	VI
5.	27	Kondapulkal	4478	2.54	VI
6.	10	Chamanpalle	3026	2.31	VI
7.	21	Dwshedshala	3428	2.31	VI
8.	24	Annaram	3189	1.98	VI
9.	5	Arepalle	2403	1.53	VI
10.	11	Charla-Boothkur	2469	1.53	VI
11.	17	Chintakunta	2789	1.53	VI
12.	20	Bommakal	3489	1.53	VI
		SULTANABAD			
1.	52	Sultanabad	7105	38.49	II
2.	3	Kapoaravpet	2051	24.82	II
3.	29	Bommareaddypally	5084	17.84	III
4.	8	Endapalle	4092	13.19	III
5.	34	Pathepalle	2158	12.67	IV
6.	65	Garrepalle	4345	12.01	IV

(Contd.)

TABLE 4.8 : *Contd.*

1	2	3	4	5	6
7.	22	Rachepalle	2218	12.03	IV
8.	58	Elgaid	4152	8.06	IV
9.	42	Wadkapuram	4003	5.51	V
10.	43	Jutapalle	5237	5.31	V
11.	72	Kolanoor	6318	5.24	V
12.	26	Myadaram	5941	5.11	V
13.	76	Kangorthi	4786	4.33	V
14.	45	Teelkunta	4057	2.96	VI
15.	15	Chegaon	2452	2.37	VI
16.	46	Dhoolkatta	2642	2.37	VI
17.	31	Ellapuram	2296	2.05	VI
18.	9	Pathagudur	2096	1.85	VI
19.	36	Katkanpalle	3618	1.85	VI
20.	44	Peddapuram	2670	1.85	VI
21.	53	Poosala	2694	1.85	VI
22.	12	Godeselapet	2047	1.76	VI
23.	39	Kothur	2108	1.76	VI
24.	40	Kummarikunta	2413	1.76	VI
25.	73	Kankula	2283	1.72	VI
26.	7	Kothapet	2385	1.59	VI
27.	41	Kachapur	2405	1.5	VI
28.	27	Bothalvanporthi	2222	1.24	VI
29.	71	Regadinadkunta	3941	1.2	VI
30.	49	Kadurpals	2985	1.11	VI
		PEDDAPALLI			
1.	111	Pothkapalle	5944	35.7	II
2.	50	Kamanpuram	5192	22.17	III
3.	95	Srirampuram	4180	17.85	III
4.	99	Odela	5747	11.77	IV
5.	62	Begumpet	3701	9.56	IV
6.	40	Palakurthi	4660	9.14	IV
7.	36	Esalatakkalapalle	4402	7.4	IV
8.	91	Kunaram	3063	7.4	IV
9.	45	Appannapet	4754	4.98	V
10.	85	Mulasala	2045	4.07	V

1	2	3	4	5	6
11.	46	Raghavapur	4056	3.55	VI
12.	41	Paltnarm	2903	3.03	VI
13.	66	Kalvacherla	2676	3.03	VI
14.	90	Pegadapalle	3473	2.83	VI
15.	43	Ranavd	2095	2.61	VI
16.	96	Gangaram	2690	2.61	VI
17.	9	Murmoor	2325	2.37	VI
18.	87	Vennampalle	2999	2.31	VI
19.	13	Kaidardi	2721	1.69	VI
20.	21	Jallaram	5057	1.53	VI
21.	51	Julapalle	2800	1.53	VI
22.	86	Maredgonda	2955	1.37	VI
23.	84	Rampalle	3874	1.27	VI
24.	48	Ranpikunta	2012	1.27	VI
25.	2	Kukkalagudur	2484	1.11	VI
26.	31	Kannala	2454	1.11	VI
27.	74	Turkalamaddikunta	2638	1.11	VI
28.	16	Madupalle	3241	1.11	VI
29.	100	Komera	2050	1.11	VI
30.	110	Chanagonda	2428	1.11	VI
31.	11	Goyalwada	2958	1.04	VI
32.	19	Madupalle	2301	0.59	VI
33.	25	Maredupaka	5117	0.52	VI
34.	33	Mallialpalle	2836	0.52	VI
35.	27	Elkalpalle	3929	0.26	VI
		MANTHANI			
1.	49	Mutharam	3064	12.9	IV
2.	3	Gunjapadiga	2867	4.59	V
3.	52	Oded	2237	3.81	VI
4.	39	Khammampalle	2227	3.13	VI
5.	25	Eklaspur	2224	1.79	VI
6.	53	Srirampuram	2871	1.53	VI

(Contd.)

TABLE 4.8 : *Contd.*

1	2	3	4	5	6
		BHEEMADEVARAPALLI			
1.	3	Veldo	4497	18.42	III
2.	25	Thadkal	3388	4.59	V
3.	26	Molangur	4111	4.33	V
4.	31	Keshavapatnam	4100	4.33	V
5.	6	Uttoor	3296	3.46	VI
6.	8	Challor	3535	2.96	VI
7.	9	Korkal	4516	2.41	VI
8.	5	Pachchunoor	2962	2.31	VI
9.	12	Ghanumukkala	3624	2.18	VI
10.	14	Elbak	3943	2.18	VI
11.	7	Marnidalpelle	2502	1.53	VI
12.	4	Vegurpalle	2041	1.2	VI
13.	30	Kothaghat	2542	1.2	VI
14.	10	Reddipalle	2094	1.11	VI
15.	22	Kannapur	2070	1.11	VI
		HUZURABAD			
1.	17	Jammikunta	13883	53.86	I
2.	14	Veenavanka	6492	11.84	IV
3.	49	Keshwapur	3104	10.79	IV
4.	42	Kamalapur	7179	8.45	IV
5.	35	Uppal	5315	8.06	IV
6.	26	Chelpur	6477	7.01	IV
7.	39	Marripalligudem	3792	4.53	V
8.	5	Vavilal	5289	.26	V
9.	13	Korapalle	3901	4.26	V
10.	4	Thanugal	3548	4.2	V
11.	20	Illanthakunta	4466	3.61	VI
12.	7	Rachapalle	2886	3.48	VI
13.	41	Vangappalle	3977	3.48	VI
14.	47	Anibal	4609	3.48	VI
15.	54	Dandepalle	3009	2.96	VI
16.	48	Geedoor	2976	2.96	VI
17.	23	Mallial	3519	2.83	VI
18.	25	Madipalle	3474	2.83	VI

1	2	3	4	5	6
19.	8	Bujnoor	3008	2.83	VI
20.	2	Porhareddipalle	2168	2.77	VI
21.	28	Sirsapalle	2275	2.31	VI
22.	18	Valbapoor	2428	2.18	VI
23.	12	Bijgir Shareet	406	2.05	VI
24.	21	Sirsed	3801	2.05	VI
25.	37	Kannur	2503	2.05	VI
26.	19	Dgarmaram	182	1.63	VI
27.	27	Pothareddipet	2879	1.63	Vi
28.	16	Bethgal	2424	1.53	VI
29.	51	Kaniparthi	2536	1.53	VI
30.	31	Kanougal	3036	1.53	VI
31.	56	Suraram	2662	1.53	VI
32.	1	Kandapak	3727	1.11	VI
33.	36	Bheemapalle	2431	1.11	VI
		CHITYAL			
1.	43	Mogullapalle	3623	9.87	IV
2.	37	Akanpalle	2178	8.94	IV
3.	38	Parlapalle	2282	1.72	VI
		PARKAL			
1.	19	Motlapalle	2016	14.54	III
2.	58	Shyampet	5632	14.01	III
3.	15	Issipet	3714	2.88	VI
4.	26	Raiparthi	4696	2.25	VI
5.	23	Cherlapalle	2848	2.16	VI
6.	17	Pradoilla	2445	1.63	VI
7.	13	Nagaram	3562	1.56	VI
8.	3	Rangapuram	2763	1.53	VI
9.	22	Narlapur	2639	1.2	VI
10.	43	Kanatatmabur	30386	1.2	VI
11.	21	Vemulapalle	45	1.11	VI
12.	24	Varikole	3065	1.11	VI
13.	40	Nadikuda	2808	1.04	VI
14.	39	Vemulapalle	2966	0.68	VI

(Contd.)

TABLE 4.8 : *Concld*

1	2	3	4	5	6
15.	48	Pocharam	2311	0.59	VI
		WARANGAL			
1.	19	Oorgonda	4310	10.88	IV
2.	37	Mongauda	2207	7.08	IV
3.	23	Marcheria	2965	4.63	V
4.	22	Siddapur	3237	4.53	V
5.	7	Nagaram	3230	3.75	VI
6.	25	Chintaghat	5423	3.75	VI
7.	10	Mallareddypet	2236	2.7	VI
8.	13	Pulkurthi	2690	2.05	VI
9.	17	Atmakur	6853	1.63	VI
10.	11	Kogilvai	2129	0.26	VI

Godavari Khani, Jagtial, Koratla, Metpalli, Sriramsagar, Ramagundam, Peddapalli, Manthani, Kothapalli, Huzurabad, Lakshettipet, and Pochampad left flank colony. Of the above urban settlements of command area, Karimnagar is district headquarters, Godavari Khani and Ramagundam are mining and industrial towns, Sriramsagar and Pochampad left bank colony are the creation of Sriramsagar Project construction. The rest, excepting Koratla, are also mandal headquarters. All the urban centres of command area are growing at a faster rate which is a predominant feature of the emerging vibrant economy of the command area.

An overview of settlement phenomenon stresses the need to have an objective planning strategy to enable the process of regional development more even and vibrant. An examination of the levels of settlements potentialities underlines the essentiality to strengthen the settlement levels of III, IV, V by commissioning more services in them to enhance the settlement accessibility by fulfilling the rising needs of the masses, instead of developing the already developed ones.

FIVE

Social Setting

ANY developmental work should aim at the betterment of the society and the people who constitute that society. Provision of physical facilities and economic infrastructure for development becomes meaningless unless the social implications of such planning measures are not rationally considered. Therefore, it is essential to aim at simultaneous implementation of physical, economic and social developmental schemes. Provision of irrigation facilities is primarily meant to enhance the agricultural productivity levels in the command area in order to uplift the people living in it by making them available the newly accrued economic benefits of such a provision. However, the final objective should be social development of command area. It is, therefore, imperative on the part of planning authorities to over-see that physical and economic developmental schemes result simultaneously in improving the social well-being of the people. In the absence of simultaneous and concurrent social planning, there exists a possibility of economic benefits being cornered by a few sections of the society. Such an act deprives the benefits to rest of the society and paves the way for that segment of the society becoming vulnerable and susceptible to all ill-effects and consequences of social inequalities and social friction. The present chapter aims at presenting a picture of social setting of Sriramsagar Command Area as existed before and after the release of waters. Although, the study is, to some extent, can be termed as a base-line study; it also presents an account of the social composition of the command area's rural community, the changes in the caste composition and the relative importance of identifying the castes or caste groups which have been benefited as a result of the provision of irrigation facilities. Finally, water resource has been examined as an instrument of ushering in social change and increasing the political consciousness. A sample survey of the villages located in Sriramsagar Command Area was conducted in order to collect micro-level data required for the analysis of the social aspects detailed above.

Multi-stage stratified random sampling method with stratification without replacement was adopted in picking up the sample villages. The complete Sriramsagar Command Area except the area on northern side of the river was considered at the first stage of sampling. At the second stage all the taluks in the command area were considered. The final stratum consisted of the villages belonging to different population size classes in each of the taluks. The purpose was to maintain homogeneity in the final stratum. A sample of thirty seven villages (*n*=37) was picked up from each of the strata by means of simple random sampling using remainder approach method without replacement. The advantage of this method is that once a sample is picked, the same sample being picked up at the second stage is totally ignored. The sampling procedure involved the following steps:

1. Defining the population (*N* Serial numbers are given to size class.)
2. Selecting the random number *r* from the tables that is less than or equal to '*N*'.
3. Dividing the Random number '*R*' by '*N*' (Stratum number with respect of each size class) and noting down the remainder (1). The value corresponding to *i*th serial number gives sample value or sample village.
4. Repeating the same procedure *n* times to get sample of size *n*.
5. Taking (*n*+1) the term if number is repeated any time.

Multi-stage random sampling technique has been preferred, as the method is more representatives. In an unstratified random sampling, some stratum may be over represented, while some may be excluded altogether. Stratified random sampling ensures only desired representation in the sample of the various strata in the population. It over-rules the possibility of essential group of the population being completely excluded in the sample. The technique has greater accuracy. Stratified sampling enables us to obtain the result of know precision for each of the stratum. Stratified sample is more concentrated geographically. Accordingly, the time and money involved in collecting the data and interviewing the individuals may be considerably reduced and the supervision of the fieldwork could be allotted with greater ease and convenience. Stratified sampling presupposes the knowledge of the strata sizes in the total number of sampling units in each stratum for the selection of the sample for each stratum. In the homogeneous stratum of the size class, 5 per cent sample is equally superior than with 10 per cent or more.

Having picked out the required sample, a questionnaire was designed to elicit the necessary information. The questionnaire has been so structured in such a manner as to be able to obtain the information with respect to a number of economic and social attributes of rural households. Household-wise information was collected by interviewing the village authorities and other knowledgeable persons living in the villages. Cross checks were made to keep a strict vigil on the authenticity of information given by the respondents at every step of data collection. The task was herculean, but the rewads outflowing from such a task were found to be adequate. A survey on household basis was contemplated, but was given up at the beginning stage itself, since, the amount of work in terms of man-power requirements and financial resources would not have permitted us in taking up such a stupendous task. A sample household survey would have certainly enhanced the status of the study, but it had to be given up because of the limitations. It should, therefore, be made clear that the study does not claim to be all pervading and the most comprehensive study of the social setting of Sriramsagar Command Area. Only certain aspects of the society in the command area which were thought to possess importance in the context of economic development were analysed.

Caste Composition

Indian society has been functioning on caste basis since times immemorial. This was perhaps required in a tradition bound Hindu society for the smooth conduct and distinct demarcation of various economic and social functions

TABLE 5.1 : **List of Sample Villages : Taluk-wise and Size Class-wise**

Sl. No.	Taluk	Population Size Class				
		0-1000	1001-2000	2001-3000	3001-5000	Above 5000
1.	Metpalli	Gorrapalli	Sirpur	Mogalpet	Maidipalli	Ibrahimpatnam
2,	Jagtial	Konapuram	Thimmapuram	Jaina	Bathkapalle	Rajaram
3.	Peddapalli	Muktaraopet	Godiselpet	Maredupake	Rampalli	Pothkapalli
		Chinnakalvole	Gadiselpet	Vennampalli		
		Nimmapalli	Chanagonda			
			Peddabonkur			
4.	Manthani	Bhatpalli	Maidembanda	Kamanpalli		
5.	Karimanagar	Meddikunta	Malkapur	Maqdoompur	Gangipalli	Choppadandi
6.	Huzurabad	Bonthupalli	Amdalpalli	Mamidalpalli	Challoor	Uppal
			Pothreddipalli			

that used to be performed by the populace of ancient and medieval India. The continued existence of caste-ridden society in modern and forward looking India can be said to be an anchronism. Despite the stringent economic and constitutional remedial measures taken by the government in the post-independent era, the barriers and limitations imposed by caste structure have not yet been overcome specially in the rural segment of society. This is true in the case of Sriramsagar Command Area also. Broadly speaking the society in the command area is organised mostly on caste lines. It would be, therefore, worthwhile to study the caste composition of the society in command area and to relate the same with other socio-economic phenomena.

The important castes, caste-groups that are found in Sriramsagar Command Area are Malas, Madigas (Scheduled Castes), Reddies, Tenugus, Yadavas, Gouds, Padmashalis, Artisans etc. The caste composition as such has not altered appreciably during the twenty-year period 1961-81 (Table 5.2). However, the relative importance of some of the castes underwent a sea change specially in the economic sphere. The scheduled castes of Malas and Madigas together accounted for 19.2 per cent of the total households in the sample villages in 1981. A slight decline in the proportion of their population can be noticed during the preceding twenty-year period. As was mentioned earlier, the scheduled caste population is mostly engaged as agricultural labourers in the fields owned by the people of other castes. Their distribution in the command area is negatively correlated with literate population. The irrigated sector of the study area registered a relatively higher growth of scheduled caste population than the unirrigated sector during the period 1961-81. This could be due to the intra-command area migration of agricultural labourers in the fields owned by the people of other castes. Their distribution in the command area is negatively correlated with literate population. The irrigated sector of the study area registered a relatively higher growth of scheduled caste population than the unirrigated sector during the period 1961-81. This could be due to the intra-command area migration of agricultural labourers from dry unirrigated taluks to the irrigated areas where the scope for working as agricultural labourers is much more. Field investigations revealed that the economic lot of the few scheduled caste people who

TABLE 5.2 : **Sriramsagar Command Area : Caste Composition - 1981**
Figures Indicate No. of Households in the Sample Village

Sl. No.	Village	Reddi	Telugu	Yadav	Padmashli	Muslim	Mala	Madiga	Vysyas
1	2	3	4	5	6	7	8	9	10
1.	Ibrahimpatnam	45	-	100	400	25	200	200	26
2.	Maidpalli	62	10	100	60	14	80	150	6
3.	Mogulpet	32	37	20	32	15	17	26	6
4.	Sirpur	70	-	11	8	7	15	15	-
5.	Gorrapalli	8	5	11	20	-	54	25	4
6.	Goderalpet	60	15	40	70	2	30	30	8
7.	Maredupak	70	60	50	30	6	-	30	8
8.	Rampalli	-	250	15	35	6	-	70	4
9.	Venompalli	23	75	60	10	12	30	100	10
10.	Peddabunkar	10	100	20	30	30	20	50	6
11.	Chinnakalavala	-	15	15	12	-	-	-	-
12.	Pothkapalli	100	50	110	60	40	20	100	30

13.	Nimmanpalli	14	2	6	-	-	-	36	1
14.	Chanagonda	10	-	40	6	6	5	5	-
15.	Gathepalli	-	7	-	12	40	40	40	6
16.	Mukkataraopet	-	-	-	5	-	20	30	12
17.	Mamiddipalli	100	30	50	30	3	-	100	6
18.	Challor	125	60	70	80	60	-	60	6
19.	Ponthapalli	30	-	40	15	-	30	20	1
20.	Amdalpalli	20	-	25	-	-	25	30	2
21.	Pothareddipalli	20	60	10	2	12	20	70	6
22.	Uppal	-	120	80	50	50	50	100	25
23.	Maidambande	30	100	8	10	1	25	25	5
24.	Khommpalli	30	10	15	5	1	-	50	10
25.	Bhattapalli	-	1	15	-	1	1	5	-
26.	Choppodendi	170	100	150	300	20	50	100	30
27.	Malkapur	12	15	-	10	3	40	50	-
28.	Magdumpoor	18	10	40	25	4	-	50	

(Contd.)

TABLE 5.2 : *Contd.*

1	2	3	4	5	6	7	8	9	10
29.	Ganglipalli	35	50	60	11	4	60	60	11
30.	Maddikunta	10	25	25	20	7	6	30	–
31.	Rajaram	–	8	120	57	3	–	74	35
32.	Nakkelapet	–	30	5	20	–	15	45	–
33.	Issazpalli	1	–	50	25	2	25	40	–
34.	Timmapur	–	–	40	22	2	200	35	22
35.	Laxmipuram	–		30	6	2	150	40	20
36.	Pathakapalli	156	82	–	156	13	135	105	19
37.	Jaina	55	35	4	60	46	30	45	23
	Command Area	1289	1372	1445	1794	445	1393	1191	352

Sl.No.	Village	Velema	M. Kapu	Perika	Goud	Jangam	Others	Total
1	2	11	12	13	14	15	16	17
1.	Ibrahimpatnam	2	100	-	30	300	281	1709
2.	Maidapalli	10	40	7	20	-	159	718
3.	Mogulpet	2	60	-	20	-	172	439
4.	Sirpur	3	30	2	10	-	93	264
5.	Gorrapalli	3	20	4	6	-	45	205
6.	Goderalpet	-	30	-	12	1	50	348
7.	Maredupak	-	10	-	5	11	64	344
8.	Rampalli	-	-	-	35	-	55	470
9.	Venompalli	-	-	150	40	3	76	589
10.	Peddabunkar	-	-	-	23	-	47	319
11.	Chinnakalvala	-	-	-	48	-	47	137
12.	Pothkapalli	-	20	-	60	-	212	802

(Contd.)

TABLE 5.2 : *Concld.*

1	2	11	12	13	14	15	16	17
13.	Nimmanpalli	-	40	-	-	-	14	113
14.	Chanagonda	15	-	15	45	-	30	117
15.	Gathepalli	5	15	-	-	-	89	254
16.	Mukkataraopet	10	-	60	30	-	32	199
17.	Mamiddipalli	-	14	-	10	-	34	377
18.	Challor	3	6	-	40	-	131	641
19.	Ponthapalli	-	5	-	60	-	42	243
20.	Amdalpalli	-	4	-	64	-	17	197
21.	Pothareddipalli	-	-	-	100	-	134	434
22.	Uppal	150	150	-	100	-	274	1149
23.	Maidambande	1	-	-	10	-	72	287
24.	Khommpalli	-	-	200	10	-	136	467
25.	Bhattapalli	-	1	-	1	1	1	27
26.	Choppodendi	-	11	-	150	-	332	1413
27.	Malkapur	-	15	-	20	-	35	200
28.	Magdumpoor	20	20	-	50	-	167	408

29. Ganglipalli	35	20	-	70	-	141	557
30. Maddikunta	-	-	-	60	-	23	206
31. Rajaram	2	144	-	104	2	210	769
32. Nakkelapet	-	70	-	50	-	245	380
33. Issazpalli	2	30	-	15	-	33	223
34. Timmapur	8	12	12	40	-	53	446
35. Laxmipuram		20	12	70	-	53	403
36. Pathakapalli	31	52	-	32	-	126	977
37. Jaina	16	185	-	6	-	143	658
Command Area	288	1124	462	1447	338	-	-

own their often small land holdings has improved after the provision of irrigation facilities and now they need not supplement their income by duplicating as farm labourers. This has resulted in shortage of agricultural labour, causing migration from the eastern part of command area.

The caste group which occupies the second position in numerical terms consists of farming communities of Tenugus and Munnuru Kapus. Both these communities have traditionally been engaged in agricultural activity not only in Sriramsagar Command Area, but also in the entire Telagana Region. The people belonging to these castes are perhaps the only communities whose main job has always been actual farming. Although all of them might not have been owner cultivators previously and still may not be considered as belonging to the class of big landlords; an overwhelming proportion of the people belonging to these castes now form the bulk of the marginal and small farmers owning their lands. The farming communities of Munnuru Kapu and Tenugus have great potential in the newly emerging agricultural scene of the command area. The people belonging to these castes possess the necessary skills and techniques of intensive cultivation which are so necessary in making optimum utilization of the agricultural innovations in the command area.

Gouds, Yadavas and Padmashalis also form a significant proportion of command area population. The traditional occupation of the Gouds has always been in operations relating to manufacturing and distributing of toddy and country liquor. As a rule, no other segment of the society is allowed to engage themselves in the occupation pursued by Gouds, as a result of which, they have been monopolising this monetarily important economic activity. Provision of assured irrigation facilities generated additional income in many households. The income so generated is normally spent on consumption of liquor. The fact that the growth rate of sales in liquor is more than double the population growth rate during the period 1961-81 proves our contention. The Goud community, therefore, stands to benefit because of the increased turn over of the trade. Although it may not be possible to make a generalisation that the Goud community as a whole can be branded as a rich community, it can safely be said that there has been a general improvement

TABLE 5.3 : **Sriramsagar Command Area : Caste Composition of Sample Villages Taluk-wise - 1961 to 1981**

(Figures in No. of Household)

Sl. No.	Taluk	Year	Reddi	Telugu	Yadav	P.shali	Muslim	Mala	Madiga	Vysyas	Velamas	M.Kapu	Parika	Brahmins
1	2	3	4	5	6	7	8	9	10	11	12	13	14	15
1.	Metpalli	1961	84	7	152	250	28	228	267	27	8	125	4	7
		1971	212	16	209	446	68	555	284	119	29	513	24	13
		1981	217	52	242	520	61	366	416	42	20	250	13	15
2.	Peddapalli	1961	115	317	207	164	127	113	323	44	23	67	160	7
		1971	217	407	256	205	148	134	389	74	26	80	187	9
		1981	287	514	356	270	150	165	491	85	30	115	225	10
3.	Huzurabad	1961	185	165	155	102	82	87	250	22	101	133	-	9
		1971	220	215	222	142	100	120	295	37	122	149	-	12
		1981	295	270	275	177	125	125	380	46	153	179	-	17
4.	Manthani	1961	23	69	30	10	3	16	68	12	1	1	200	-
		1971	28	89	31	10	3	21	73	13	1	1	200	-
		1981	33	111	38	15	3	26	80	15	1	1	200	-

(Contd.)

TABLE 5.3 : *Contd*

1	2	3	4	5	6	7	8	9	10	11	12	13	14	15
5.	Karimnagar	1961	196	82	170	239	26	94	160	26	50	41	–	15
		1971	222	166	210	303	32	124	210	39	55	54	–	7
		1981	245	200	275	366	38	156	240	45	55	66	–	8
6.	Jagtial	1961	158	89	169	211	42	321	240	76	20	300	11	8
		1971	175	130	211	301	53	437	304	91	23	375	17	11
		1981	212	165	259	446	68	555	384	119	29	513	24	13
	Commad Area	1961	7.67	6.98	8.46	10.31	2,95	8.23	12.53	1.98	1.94	6.39	2.59	0.34
	(% Proportion	1971	7.63	8.04	8.15	9.65	2.77	9.54	11.35	2.55	1.75	8.04	2.93	0.35
	of the caste)	1981	7.36	7.84	8.26	10.25	2.54	7.96	11.38	2.01	1.64	6.42	2.64	0.36

Sl. No.	Taluk	Year	S. Jangam	Goud	Kummari	Vaddari	Christian	Bestha	Manne
1	2	3	16	7	18	19	20	21	22
1.	Metpalli	1961	200	46	15	1	-	50	-
		1971	2	24	17	16	-	67	-
		1981	300	86	82	45	-	232	-
2.	Peddapalli	1961	12	185	63	9	2	50	3
		1971	14	221	89	12	2	55	3
		1981	15	298	109	16	3	50	3
3.	Huzurabad	1961	223	223	20	2	14	8	-
		1971	-	283	27	4	18	10	-
		1981	-	374	43	6	20	10	-
4.	Manthani	1961	1	16	5	1	-	20	-
		1971	1	16	5	1	3	20	-
		1981	1	22	5	1	-	25	-

(Contd.)

TABLE 5.3 : *Contd.*

1	2	3	16	17	18	19	20	21	22
5.	Karimnagar	1961	-	195	48	3	20	25	16
		1971	-	225	65	3	25	30	31
		1981	-	350	85	3	30	40	51
6.	Jagtial	1961	1	205	9	1	1	38	19
		1971	1	324	13	1	2	58	25
		1981	2	317	15	1	3	79	35
	Command Area	1961	2.05	8.34	1.53	0.16	0.35	1.83	0.36
	(% Proportion	1971	0.12	7.49	1.48	0.25	0.34	1.64	0.40
	of the caste)	1981	1.81	8.27	1.93	0.41	0.32	2.54	0.50

Sl. No.	Taluk	Year	Erakala	Mangli	Sakali	Tailor	Artisans	Nethkani	Others
1	2	3	23	24	25	26	27	28	29
1.	Metpalli	1961	1	11	13	5	29	5	59
		1971	1	19	27	10	47	20	113
		1981	3	39	61	34	68	-	147
2.	Peddapalli	1961	6	24	98	15	136	-	15
		1971	7	29	114	18	135	-	155
		1981	7	35	130	24	148	24	181
3.	Huzurabad	1961	9	8	63	22	76	-	95
		1971	11	10	42	30	98	-	321
		1981	15	15	112	38	136	-	233
4.	Manthani	1961	1	2	16	5	23	90	3
		1971	2	2	16	5	24	110	3
		1981	3	2	21	7	26	143	3

(Contd.)

TABLE 5.3 : *Concld.*

1	2	3	23	4	25	26	27	28	29
5.	Karimnagar	1961	7	6	52	7	104	20	57
		1971	11	9	72	7	242	30	74
		1981	13	13	102	8	178	40	103
6.	Jagtial	1961	2	28	100	6	61	80	82
		1971	2	37	125	6	87	95	102
		1981	2	54	172	8	116	115	128
	Command Area	1961	0.24	0.75	3.27	0.57	4.11	1.96	2.98
	(% Proportion	1971	0.23	0.72	2.71	0.52	4.32	1.74	5.40
	of the caste)	1981	0.24	0.90	3.41	0.68	3.84	1.84	4.54

Source : Sample Survey of Command Area Villages.

in the economic and social status of the people belonging to Goud caste.

The proportion of population belonging to Yadava (shepherd) community is significant figure-wise and is near equal to the shares of Gouds, Tenugus and Munnuru Kapus. However, the community's lot has not improved in the recent decades due to a number of factors. Foremost of these factors is the decrease in pasture and grazing lands by invasion of cultivation. More and more agriculturally waste lands and pasture lands are being brought under the plough causing a shrinkage of the traditional occupation base of the shepherd class. Provision of irrigation facilities did not bring any prosperity to Yadavas as it did in case of farmer castes and Gouds. However, that section of the Yadava community which had the side occupation of farming along with the traditional activity of sheep grazing, was benefited by the provision of irrigation facilities.

Reddies, Velamas and to some extent Vysyas once formed the social and economic elite of the command area. The first two castes can better be termed as the land owning caste, as against the land tilling castes of Munnuru Kapu, Tenugu and the scheduled castes. Reddies and Velamas were perhaps the first beneficiaries of the agrarian changes. However, the people belonging to land tilling castes started asserting their economic and social rights. They were aided in their efforts by stringent land tenure and land ceilling enactments made by the Government. As a result of this, the supermacy of Reddies and Velamas was broken up to some extent, at least, in some pockets of the command area. Even then, Reddies and Velamas are continuing to assert themselves, though to a lesser degree, in the political, economic and social scenario of the command area. They are still a force to reckon with. Vysyas who were once the only people with a monopoly in trade coupled with money-lending, have lost the grip on their strongholds as a result of many new factors which cropped up in recent years. Nationalisation of scheduled banks and consequent expansion of the banking operations in the rural areas and increased lending by the agencies of State and Central Governments helped the process.

The traditional occupation of Padmashalis has been weaving of textiles. Of late, people belonging to Padmashali caste had to change their occupation from cloth weaving to other economic activities. This was because of the decreasing demand for hand woven cloths in urban as well as rural areas. Field investigations revealed that the Padmashalis have branched out to take up farming, petty trade etc.

Telangana of which Sriramsagar Command Area is a part, has traditionally been the area of the well known Bhawanti type of house. It is still the most dominant house type in rural Telangana, though a number of minor innovations in its design and roof were made in recent decades. The main features of this house is a central courtyard surrounded by living rooms and a roof made of country tiles. However, some families whose economic position showed an upward movement, are opting for Mangalore tiled houses and reinforced cement concrete houses. The R.C.C. and Mangalore Tiled (M.T.) houses are relatively more expensive to build and have thus become symbols of economic and social status.

Four distinct types of houses which form an overwhelming proportion of houses in Sriramsagar Command Area are identified. They are :

First Order : Reinforced cement concrete house with bricks walls and steel and cement roof.

Second Order : Houses with M.T. or corrugated iron sheets roof with walls made up of either mud or bricks.

Third Order : Bhawanti type of houses with country tile roof with mud or brick stone walls (C.T.).

Fourth Order : Thatched roof houses with mud walls. Thatched walls are uncommon (T.R.).

If we consider Sriramsagar Command Area as a whole, more than half of the total houses in 1981 belonged to the third category of houses with country tiles. A little less than one-third of the total houses are still constructed out of thatch material (Table 5.4). Mangalore tiled houses come next with 10.3 per cent. The R.C.C. roofed houses are still rare with a share of just 2.6 per cent. However, there has been a significant improvement in the proportion of houses belonging to first and second order houses over the period 1971-81. It is interest-

TABLE 5.4 : Sriramsagar Command Area : Caste-wise Distribution of House Types - 1971-81

Caste Group	Year	House Type									
		R.C.C.		M.T.		C.T.		T.H.		Total	
		No.	%	No.	%	No.	%	No.	%	No.	%
1	2	3	4	5	6	7	8	9	10	11	12
Reddy	1971	20	2.4	35	4.3	690	85.3	65	8.0	810	100.0
	1981	89	8.5	199	18.5	699	66.6	57	5.4	1044	100.0
Velama	1971	1	0.4	2	0.8	198	87.2	26	11.6	227	100.0
	1981	14	4.8	39	13.3	226	77.4	13	4.5	292	100.0
Vysya	1971	8	5.9	8	5.9	118	88.2	-	-	134	100.0
	1981	30	17.9	16	9.5	118	70.2	4	2.4	168	100.0
Madiga	1971	-	-	-	-	205	25.4	603	74.6	808	100.0
	1981	6	0.5	190	14.6	414	31.9	688	53.0	1298	100.0
Mala	1971	-	-	4	1.1	108	28.5	266	70.4	378	100.0
	1981	2	0.3	24	4.3	253	45.6	276	49.8	555	100.0

(Contd.)

TABLE 5.4 : *Contd.*

1	2	3	4	5	6	7	8	9	10	11	12
Tenugu	1971	-	-	-	-	331	40.7	482	59.3	813	100.0
	1981	4	0.3	89	7.3	628	51.3	503	41.1	1224	100.0
Munnuru Kapu	1971	-	-	1	0.3	221	68.0	102	31.7	325	100.0
	1981	17	3.2	34	6.4	381	71.3	102	19.1	534	100.0
Yadava	1971	-	-	-	-	402	51.1	384	48.9	786	100.0
	1981	5	0.4	58	5.0	822	74.8	214	19.8	1099	100.0
Total	1971	29	0.7	50	1.2	2273	53.1	1928	45.0	4280	100.0
	1981	167	2.6	649	10.3	541	36.3	1937	30.8	6294	100.0

Note : Only Important Castes Included.

R.C.C. : Reinfrced Cement Concrete Roofed House. M.T. : Mangalore Tile House

C.T. : Country Tiled House.

T.H. : Thatched Roof House

Source : Sample Survey of India.

ing to note that the number of R.C.C. houses went up by more than five times during the period, whereas, there was insignificant increase in the share of country tiled houses and a decline with respect to thatched houses. It can, therefore, be safely assumed that there has been a discernible transformation of house types probably as a result of increased incomes of the families inhabiting the area. Decrease in the proportion of thached houses, which are considered to be not very convenient and comfortable for living, is really a very welcome development in the command area. The general improvement in housing conditions, though uninversal in the area, certain trends can be deciphered from Tables 5.4 and 5.5, which show caste-wise and taluk-wise distribution of house types. It can be observed from the tables that there has been an upward shift of people belonging to all sections of the society, from the lower order house types to high order house types. As many as 85 per cent of the forward communities of Reddies and Velamas were living in the third order houses in 1971; as against this, only 66.6 per cent of Reddies and 77.4 per cent of Velmas were living in third order settlements in 1981 (country tiled houses). The decline in the share during the decade can be explained by the fact that there has a significant improvement in the proportion of Reddy and Velama families dwellings belonging to the first order (R.C.C. houses) and the second order (Mangalore tiled houses). It can be inferred that there has been a significant improvement in the economic condition of these caste people facilitating them to opt for the construction of higher order houses. The improvement in housing conditions is not just confined to the people belonging to forward castes only. There has been a shift to the higher order in middle rung farming communities and scheduled castes also. A major proportion (59.3 per cent) of Tenugus were living in thatched houses in 1971; whereas, in 1981, only 41.1 per cent of the Tenugus were living in thatched houses. The proportion of people living in the third order house type has gone up from 40.7 per cent in 1971 to 51.3 per cent in 1981. Similarly, the scheduled caste families are now increasingly going for country tiles houses as is evident by an increase in the proportion of the third order house type and decrease in the proportion of fourth order house type. It is not out of place to observe that not a single

TABLE 5.5 : **Sriramsagar Command Area : Caste and House Types in Sample Villages (Taluk-wise) 1971–81**

Taluk		Reddy				Velama			
		R.C.C.	M.T.	C.T.	T.H.	R.C.C.	M.T.	C.T.	T.H.
Metpalli	1971	6	8	155	2	–	–	9	–
	1981	57	11	72	–	6	7	3	–
Jagtial	1971	–	–	39	–	–	–	12	–
	1981	–	16	40	–	–	2	16	–
Peddapalli	1971	2	11	201	45	1	2	29	–
	1981	11	78	174	42	4	12	34	–
Mantham	1971	–	–	20	1	–	–	–	–
	1981	–	1	27	5	–	–	–	–
Karimnagar	1971	10	14	114	12	–	–	40	5
	1981	15	52	161	2	4	3	45	3

Huzurabad	1971	2	2	201	5	–	–	108	21
	1981	6	44	335	3	–	15	128	10
Command Area	1971	20	35	690	65	1	2	198	26
	1981	89	199	699	57	14	39	226	13

		Vysya				Madiga			
		R.C.C.	M.T.	C.T.	T.H.	R.C.C.	M.T.	C.T.	T.H.
Metpalli	1971	1	1	4	–	–	–	54	33
	1981	1	2	9	–	5	32	35	86
Jagtial	1971	1	–	15	–	–	–	2	39
	1981	5	3	19	–	1	39	10	37
Peddapalli	1971	3	5	50	–	–	–	16	315
	1981	7	4	57	4	–	6	92	262

Contd.

TABLE 5.5 : *Contd.*

Taluk		R.C.C.	M.T.	C.T.	T.H.	R.C.C.	M.T.	C.T.	T.H.
Mantham	1971	-	-	10	-	-	-	2	34
	1981	-	-	13	-	-	1	16	57
Karimnagar	1971	-	2	18	-	-	-	59	72
	1981	5	6	33	-	1	100	129	10
Huzurabad	1971	3	-	21	-	-	-	72	112
	1981	12	1	31	-	-	12	102	236
Command Area	1971	8	8	118	-	-	-	205	603
	1981	30	16	160	4	6	190	414	688
		Mala				Tenugu			
Metpalli	1971	-	-	62	46	-	-	11	10
	1981	8	14	109	26	-	-	28	7

Jagtial	1971	–	–	4	21	–	–	5	11
	1981	–	–	14	46	–	3	20	-
Peddapalli	1971	–	–	12	92	–	–	201	169
	1981	–	4	52	90	1	48	265	234
Mantham	1971	–	–	2	20	–	–	2	49
	1981	–	–	8	22	–	15	26	72
Karimnagar	1971	–	4	18	31	–	–	51	59
	1981	–	5	47	16	–	7	167	61
Huzurabad	1971	–	–	10	56	–	–	61	184
	1981	–	1	23	76	3	16	122	129

Concld.

TABLE 5.5 : *Concld.*

Taluk		R.C.C.	M.T.	C.T.	T.H.	R.C.C.	M.T.	C.T.	T.H.
Command Area	1971	-	4	108	266	-	-	331	482
	1981	2	24	253	276	4	89	628	503
		Mannuruu Kapu				Yadav			
Metpalli	1971	-	1	48	5	-	-	33	15
	1981	17	5	65	3	2	2	72	1
Jagtial	1971	-	-	41	15	-	-	29	15
	1981	-	14	59	22	1	9	40	12
Peddapalli	1971	-	-	24	38	-	-	108	158
	1981	-	9	48	53	-	11	208	112

Mantham	1971	–	–	2	2	–	–	12	12
	1981	–	1	7	3	–	5	29	10
Karimnagar	1971	–	–	24	1	–	–	108	112
	1981	–	2	48	–	2	9	239	20
Huzurabad	1971	–	–	82	41	–	–	112	72
	1981	–	3	154	21	–	22	234	59
Command Area	1971	–	1	221	102	–	–	102	384
	1981	17	34	381	102	5	58	822	214

Source : Sample Survey of Villages

R.C.C. : Reinforced Cement Concrete House, M.T. : Mangalore Tiled House, C.T. : Country Tiled House,

T.H. : Tatched Roof House

scheduled caste was living in either R.C.C. or M.T. house in 1971. In 1981, 15.1 per cent and 4.6 per cent of scheduled castes were dwelling in houses belonging to the first two high order house types. However, these figures are nowhere near those of forward castes. They have a long way to go when compared to forward castes.

The improvement in housing conditions in Sriramsagar Command Area, though ubiquitous is embedded with some regional variations as can be seen from Table 5.5. Generally speaking, degree of improvement in housing conditions is much more evident in the irrigated sector compared to unirrigated sector.

Water Resources as an Instrument of Social Change

Social change may be defined as modification in ways of living, doing and thinking of people. It refers to the modifications which occur in the life patterns of people. According to Anderson and Parker social change is the structure of functioning of societal forms or processes. Social changes are variations from the accepted or pre-existing modes of life - whether due to alteration in geographical conditions, in cultural equipment, composition of population, or ideologies and whether brought about by diffusion or inventions within the group (Gillian and Gillian).

A number of gigantic irrigation projects have come up in India with the intention of providing much needed irrigation facilities to the parched lands in the semi-arid parts of the country. The primary aim in constructing these projects is the economic upliftment of the command areas by increasing the productivity levels in the agricultural sector. However, it has been observed that the provision of irrigation facilities is also found to cause transformation of the society. It is hypothesised that the provision of water generates vibrations in the societal structure.

As mentioned earlier, the type of house in which a family lives besides reflecting the economic status, is also an indicator of the social status of that family. Bhawanti type country tile roofed house built with indigenous building materials was the most commonly seen house in the area of study before the onset of irrigation facilities. This type of house used to be the abode of most of the socially developed classes e.g., Brahmins,

Vysyas, Velamas, Reddies, etc. Most of the other castes specially the scheduled castes and tribes used to live in simple, easy to build and cheap thatched roof houses. As the benefits of irrigation facilities have started accruing, the pattern of house types has markedly changed. Although country tile roofed houses are still in existence, they have lost their predominant position. Relatively more and more modern houses built of R.C.C. and Mangalore tiles have sprung up in the country side. There is also a substantial decrease in the number of thatched roof houses. Not only that, the ownership of pucca Bhawanti-type houses and modern R.C.C. houses are no more the monopoly of forward castes. A substantial number of newly built mangalore tile roofed houses are owned by scheduled caste and backward class people.

Mobility as a social factor is dependent on the availability of transport facility, level of economic development and social awareness. Transport facility acts as an instrument to deliver the information; in the process creating interaction among the villages and towns which, in turn, infuses new ideas into the countryside. It is observed that the people were deprived access with the rest of the world for generations as there was no physical infrastructure. Irrigation gave the fillip to the country as in the case of other sectors.

Most of the villages in the command area are now well connected by motorable roads which were laid by project planners as a matter of policy. Four out of five villages in Metpalli, six out of seven in Jagtial, eight out of eleven in Peddapalli, two out of three in Manthani, three out of five in Karimnagar and five out of six sample villages in Huzurabad are provided with motorable road facility. Although some bus services were in existence before, provision of water made it feasible to have greater bus access. Bus service frequency has also increased manifold. In Metpalli, for example, it went up by four-fold and is followed by Jagtial with three times increase. Bus service is a recent phenomenon in the cases of many villages of the rest unirrigated segment of command area. Private buses operation is predominant and introduction of RTC is a recent one.

The utilities considered in this study are bicycles, motorcycles and cars. It is found that there is a spate of increase in the number of bicycles. Study also reveals (Table 5.6) that the man-bicycle ratio has been tremend-

TABLE 5.6 : **Man-Bicycles Ratio**

Taluka	1961	1962	1981
Metpalli	27	28	27
Jagtial	186	178	47
Peddapalli	115	42	21
Manthani	–	–	143
Karimnagar	355	55	9
Huzurabad	469	166	51
S C A	100	53	19

ously influenced by economic prosperity. Prior to 1971, when the water was not yet released, the man-cycle ratio was very low when compared to the sudden spurt in the ratio after the provision of assured irrigation facilities. The provision of good transportation facilities in the countryside of the command area has led to a greater degee of mobility of the people specially those belonging to hitherto deprived classes. This has resulted in greater interaction between places and people and also diffusion of ideas, habits, life styles, etc. It can be observed that as a consequence of greater degree of mobility a new set of attitudes and social relations have emerged in the changed scenario.

A more vibrant change is felt in rapid expansion of commercial establishments (Table 5.7) and their typology. Commerce in rural areas used to be the monopoly of Vysya community since generations. A striking change can be observed in the type of units and ownership pattern. Prior to the provision of water, the commercial activity was limited to *Kirana* shops only. Provision of water dismantled the age old tradition of commerce being run by Vysyas only. It led to the emergence of a different spectrum, in which the activity is no more the monopoly of Vysyas. Not only people from other castes are now engaged in trade and commerce, but many new types of commercial establishments like cloth

TABLE 5.7 : Population-Commercial Units Ratio

	Kirana	Cloth	Hotel	Pan	Cycle Taxi	Others
	1	2	3	4	5	6
				1961		
Metpalli	3881	Nil	Nil	Nil	Nil	Nil
Jagtial	1781	-	-	-	-	-
Peddapalli	1712	-	13702	-	-	-
Manthani	-	-	-	-	-	-
Karimnagar	1242	-	6211	-	-	-
Huzurabad	977	-	11929	-	-	-
Sriramsagar Command Area	1547	-	14310	-	-	-
				1971		
Metpalli	4061	Nil	-	-	-	-
Jagtial	2392	-	-	-	-	-
Peddapalli	1674	-	8372	-	8372	-
Manthani	1129	3388	-	-	-	-

(Contd.)

TABLE 5.7 : *Contd.*

	1	2	3	4	5	6
Karimnagar	828	6211	2484	-	6211	-
Huzurabad	658	-	7240	-	-	-
Sriramsagar Command Area	1190	23800	7934	-	35701	-
				1981		
Metpalli	1424	14243	2373	4747	NA	4747
Jagtial	600	16043	3208	16000	1782	8021
Peddapalli	424	2123	805	-	2595	11677
Manthani	593	4151	-	-	-	-
Karimnagar	403	2905	998	1048	1612	1905
Huzurabad	355	3403	896	17019	1418	8510
Sriramsagar Command Area	496	3302	1197	13680	2816	5040

shops, beer shops etc. have started making their appearance. There has been a shift in traditional occupations. A marked transfer is taking place by the entry of backward communities like Padmashali, Gouda, and Perika and forward community like Muslims venturing into new non-traditional occupations. All the units, which have come up recently are captured by these communities resulting competition in the trade. Ownership pattern of *Kirana* (grocery) shops was traditionally with Vysyas and to a limited extent by Padmashali, and Muslim. Hotel, cloth, Cycletaxi, Pan shops belong to Padmashali, Gouda and Muslim and their involvement is very recent. The new class of businessmen participate in business addition to agriculture. The profits in commerce range from Rs. 200 to Rs. 800 per month in most of the cases and this is an additional income supplemental to the income from agriculture. It can, therefore, be inferred that increased commercial activity is creating a new frame of social structure upon which emerging social relations are operating.

In response to the ripples sent by assured irrigation, some communities like Vysya, Padmashali, Gouda and Muslim have migrated to the area after 1970. Migration on insignificant scale also took place during 1960s. Sixty-three per cent of sample villages have a migration composition of five to fifteen families in each village. In the other twenty-seven per cent of villages, there has not been any in migration. It is found that the distance travelled by migrants is short. Some of the in-migrants are from the submerged villages of the reservoir. However, a few villages situated in the mining belt are experiencing rapid immigration. Apart from this there has been some out-migration due to unstabe political situation and extremist activity. Landlords belonging to upper castes fled the area and settled in urban centres. These landlords who mostly belong to Reddy and Velama castes, are no more as strong a force to reckon with as they used to be before in the rural economy. Their lands are in the process of being taken over by the people belonging to backward classes and scheduled class either on tenant basis or by out right purchase.

Political scenario is like smouldering caldera. There has been a mass upsurge in the extremist activities and annihilation of feudal lords by extremists in retaliation of suppression of backward class and scheduled class people. One of the causes for the struggle seems to be the commissioning of irrigation project (Balagopal, 1983) which increased the 'commoditization'. Confrontation erupted between landlords and labourers as land has become more productive. It can also be observed that the growth of economy and monopolisation of resources by a small segment of the society polarised the class interests. Besides there is a disproprotionate balance between wages and production. Taking advantage of these conditions extremist parties took hold of the situation and organised the mass revolt against the landlords. There have been some incidents of stopping land transactions made by the landlords.

Apart form this, ubiquitous availability of water is also causing changes in the dilution of rigid traditional relationships. It is observed that the untouchables, backward castes and forward castes use the same body of the water for drinking, bathing and washing. Hitherto, before the construction of canals, water was available at point locations which intrinsically led to the seggregation of the facility. Now, since the water is available in a linear pattern, it is no more possible to have castewise seggregation. All caste people have to obtain water from the same source, thus leading to the breakdown of social barriers to some extent.

Income generated from the economic affluence of the area has been siphoned off into urban centres, leading to an alarming growth of urban centres. Emergence of strong urban centres heralded a new era of social life and acquiring urban life styles. Cinema theatres have mushroomed, buzzing trade has taken roots, and new political power centres have emerged. These factors together led to the emergence of a new class of people. The emerging structure may lead to erosion of some of the traditional relations but may lead in establishing a new social order. However, the extent of social change, its intensity, spread and consequences are yet to be assessed and probed in detail.

REFERENCES

Amarnath K. Menon, *India Today*, May 15, 1985.

Anderson and Parker, *Society*, pp. 385, 1964.

Balagopal K.A.P : Peasant Struggle and Repression in Peddapalli, *Economic and Political Weekly*. No. 20, May 15, 1982.
Gillian and Gillian, *Cultural Sociology*, p. 561, 1950.
Srinivas M.N., *Caste in Modern India and other Essays*, Asia Publishing Home, Bombay, 1962.
Srinivas M.N., *Social Change in Modern India*, Allied Publishers, New Delhi, 1966.

SIX

Levels of Development and Backwardness

A major objective of economic and physical planning measures, like commissioning of major irrigation projects is the goal of achieving social and economic justice. In order to fully understand and gauge the benefits arising out of the implementation of economic and physical plans, it is essential to identify the regions or areas according to their levels of development, typology and dimensions of backwardness. The present chapter is, therefore, aims at qualitatively identifying the levels of development or backwardness in different sub-units of Sriramsagar Command Area. Measurement of differential levels of development and backwardness in different sub-units of Sriramsagar Command Area. Measurement of differential levels of development and backwardness is necessary for strategy formulation for the development of backward areal pockets. Required policies, programmes and plans can be framed based on the specific area characteristics their requirements and capabilities.

Mandals are taken as the basic units of the present study. They have been preferred to taluks since the latter are much larger and not always homogeneous with regard to social and economic characteristics. Sriramsagar Command Area is composed of forty-six mandals spread over four districts.

The utility of any study depends to a great extent on the proper selection of the right variables. Keeping this in view the following twenty-one development indicators were selected. These indicators or variables fall in three groups.

1. Population Variables

1. Density of Population

2. Per cent proportion of scheduled caste population
3. Per cent proportion of scheduled tribes population
4. Per cent proportion of working population to total population
5. Literacy ratio
6. Population in secondary activities as per cent of total production.

2. Land-use and Agricultural Variables

7. Per cent proportion of agricultural population
8. Per cent proportion of forest area to total cropped area
9. Net cultivated areas as per cent of total area
10. Irrigated area as per cent of total cultivated area
11. Double cropped area as a per cent of net area sown
12. Cropped area as per cent of total area
13. Area under pastures as per cent of total area
14. Fallow lands as per cent of total area
15. Areas under commercial crops
16. Canal irrigated area as per cent of total irrigated area.

3. Amenities and Infrastructure Variables

17. Electrified villages as per cent of total villages
18. Road length per ten thousand population
19. Number of doctors per ten thousand population
20. Number of hospital beds per ten thousand population
21. Villages with high schools as per cent of total villages.

Data non-availability prevented consideration of number of other variables which would have been greatly helped in making the studies more broad based. Sriramsagar Command Area is mainly an agricultural entity as such greater reliance on agricultural variables is not only necessary but inevitable.

Although all the selected variables have bearing on the development/backwardness of the region, it cannot be taken for granted that the amount of importance of the above indicators is in no way equal or at least near equal. One method commonly used by many researches is the design of a suitable composite and weighted index of development/backwardness which to some extent is a reflection of the indicators. The over all pictures that can be obtained by this, gives a relatively good picture of the level of development. However, this method though statistically less cumbersome and relatively easy has in-built flaws. One such flaw is the assigning of weights which in most of the cases are given arbitrarily making from one individual researcher to another based on their value judgements. A number of methods of assigning weights and computation of weighted composite index exist. MC Granchar *et al.* (1970), Nanjappa, Rao and Simhadri (1985), Rao (1981), Drewnowski (1970) and Ashok Mitra (1961), Iyenger and Sudershan (1980) are some of the researchers who devised different methods of weighted composite indices of variables. It is not proposed to discuss these methods here since these methods of data amalgamation are considered to possess, at least, some traces of subjectivity. Keeping this in view, the levels of development and backwardness in Sriramsagar Command Area are identified by using the sophisticated statistical technique of principal component analysis, a method of factor analysis. The technique provides factors loading for each variable and these loadings are nothing but the co-efficient of correlation between the observed variables and the unknown derived factors.

Principal Component Analysis

"Principal component analysis is a method of analysing the set observations from their interrelations to determine whether the variations represented can be accounted for adequately by a number of basic categories smaller than that which the investigation was started (Fruchter, 1967). Thus, data obtained from a large number of priori measures may be explained in terms of a small number of reference variables. By adopting principal component analysis technique. However, for a given matrix of correlations, the raw factoring methods arbitrarily located the reference axes in a different position. In order to move the axes from arbitrary location to some position useful for interpretation of the factors and for comparison with other studies, axes need to be rotated. A major goal of rotation is to obtain meaningful factors that are as consistent (or invariant) as possible from analysis to analysis (Fruchter, 1967). There are various methods of rotating the factor matrix. The method used in the study is varimax rotation method.

In methematical terms, principal component model may be put as follows (Rao, H., 1984):

$$Z_j = a_{j_1} P_1 + a_{j_2} P_2 + \ldots a_{j_m} P_m$$

where

$$Z_j^{\bullet} = \frac{x_j - \bar{x}}{0 \times j} \quad \text{on standardised variable}$$

$P_i = (i=1 \ 2 \ . \ . \ . \ m)$ are the principal components

$a_{j_i} = j=1, 2 \ . \ . \ . \ n$ *are* the co-efficients or factor loadings of $i = 1, 2 \ . \ . \ . \ m$ the *j*th variable relating to *i*th component.

In other words, each principal component or factor as it is called in the present study, is nothing but a linear combination of weighted variables which can also be expressed as :

$$P_i = E^n a_j \times j$$

where

a_j = factor loading of j variables

$$j = 1, 2, \ldots n$$

or $P_i = a_{1i} \times 1 + a_{2i} \times 2 \ . \ . \ . \ a_{ni} \times n.$

It can be used to delineate homogeneous regions and to identify typological dimensions the total variance is explained in terms of *m* component on factor (Rao, H., 1984).

Factors and Eigen Values

The results of principal component analysis of Sriramsagar Command Area indicate that there are seven dimensions which together explain development/backwardness of the area. The eigen values and the cumulative percentage of the variance explained are given below. It may be mentioned here that a maximisation procedure under certain side conditions takes on a very specific and recognizable form, namely, finding eigen values of given matrix,

It is evident from Table 6.1 that the first factor alone explains as much as a quarter of total cumulative per cent variance. The first three factors combinedly account for more that half of the total variance. All the seven factors together explain factors and significant associations with in them are listed below.

TABLE 6.1 : **Eigen Values of Factors**

	Eigen Values	Cumulative % of Eigen Value
Factor - I	5.40355	25.731
Factor - II	3.25560	41.234
Factor - III	3.09200	55.918
Factor - IV	1.92581	65 128
Factor - V	1.27115	71.181
Factor - VI	1.07693	76.310
Factor VII	1.00531	81.097

Factor I (Agricultural Dimension)

In an area dominated by vibrant agricultural economy, it is but natural that the Factor one should reflect the

Factor I

	Indicator	Rotated Factor Loading
1.	Cropped area as per cent of Total area	+ 0.92
2.	Net area sown as percentage of Total area	+ 0.92
3.	Road length per 100 sq.km.	+ 0.58
4.	Proportion of villages with high schools	+ 0.43
5.	Per cent electrified villages	+ 0.43
6.	% Forest area	- 0.75
7.	% Scheduled tribe population	- 0.67

agricultural dimension. The first dimension highlights the association between cropped area, net cropped area, road length per hundred sq.km, per cent villages with high schools and per cent electrified villages. It brings out significant negative association with forest area and the proportion of scheduled tribes population.

The first factor which explains for the maximum variance, highlights the importance of agricultural activity as exemplified by high factor loading for cropped area, net area sown and the electrified villages.

The agricultural activity in turn is associated by high road length ratio and high proportion of high schools. Obviously, the forest area, where most of the scheduled tribe population lives, possesses negative factor loading.

Factor II (Urban Dimension)

The second factor reflects the association between the characteristics aspects typical of urban and semi-urban areas. They are: literacy ratio, proportion of population in secondary activities, density of population, road length ratio and hospital beds per 1,000 population.

Factor II

	Indicator	Rotated Factor Loading
1.	Literacy ratio	+ 0.89
2.	Proportion of population in secondary activities	+ 0.80
3.	Density of population	- 0.71
4.	Road-length ratio	+ 0.54
5.	Hospital Beds/Population ratio	+ 0.46
6.	Proportion of workers to total population	- 0.80
7.	Proportion of agricultural workers	- 0.79

It can be deciphered from the above that the second factor reveals the urban dimension of development/backwardness of Sriramsagar Command Area. Then high magnitude of positive factor loadings is indicative of this. The workers ratio and proportion of agricultural workers have negative factor loadings.

TABLE 6.2 : **Indicators of Levels of Development Correlation Matrix**

		1	2	3	4	5	6	7	8	9	10	11
1.	Density of Population	1.00000										
2.	% proportion of scheduled caste population	-.18403	1.00									
3.	% proportion of scheduled Tribe population	-.46196	.37659	1.00								
4.	Litracy ratio	.81598	-.17429	-.34705	1.00							
5.	Per cent proportion of workers	-.50833	-.17429	.14913	-.72500	1.00						
6.	Percent proportion of Agricultural workers	-.33503	-.17032	-.00938	-.53851	.53655	1.00					
7.	Per cent proportion of forest area	-.51778	.22625	.74309	-.21619	.09744	-.02720	1.00				
8.	Per cent proportion of net cultivated area	.27104	-.07826	-.51480	-.00101	.01670	.13791	-.60912	1.00			
9.	Per cent proportion of Irrigated area	.10918	-.17317	-.05829	.03558	.25488	-.07418	-.02037	.03476	1.00		
10.	Per cent proportion of Electrified villages	.10480	.01328	-.12038	-.10106	.14644	.14882	-.26716	.44638	-.15994	1.00	

11. Road length per 100 sq. km.	.50083	-.02642	-.02642	-.43888	.41512	-.37812	-.37467	-.44491	.44638	-.15994	1.00
12. No. of Doctors per 10,000 population	.26615	-.06241	-.15302	.23228	-.13929	.09306	-.26454	-.12032	.12852	-.49298	.08874
13. Hospital Beds per 10,000 population	.45574	-.05450	-.06692	.55587	-.32041	-.17448	-.11751	-.03970	.21572	.11161	.27729
14. % Proportion of canal irrigated area	-.25175	-.12272	.13639	-.22901	.25864	.08370	.32310	.16184	.50460	.02500	-.00992
15. % Proportion of double cropped area	.24249	-.22719	-.32318	-.12119	.14613	.30784	-.34803	.13322	.36781	-.22397	.01415
16. % prop. of crop. area	.91916	-.09513	-.53591	.06528	-.02678	.28591	.52616	.86867	.12464	.33824	.43570
17. % proportion of Pasture land	.01527	-.03383	-.08203	.01861	-.02131	-.06939	-.13773	.06499	.08704	.06908	-.02557
18. % proportion of population in secondary activities	.64334	.15127	-.26490	.71689	-.52257	-.57911	-.17538	-.00669	.02828	.14172	.38308
19. % proportion of follow lands	.27665	-.20468	-.25006	.20830	-.07088	-.01048	-.43943	-.27846	.05282	-.20 19	.04304
20. % proportion of commercial crops	.20401	-.14322	-.19921	.07527	-.03539	.13868	-.43778	-.07825	-.14047	-.20396	.07671
21. % prop. of vill.H.School	.42725	-.20769	-.45677	.33604	-.22829	-.09220	-.48084	.23544	-.01646	.25284	.34466

(Contd.)

TABLE 6.2 : *Contd.*

	12	13	14	15	16	17	18	19	20	21
1. Density of Population										
2. % Proportion of Scheduled Caste Population										
3. % proportion of scheduled Tribe population										
4. Litracy ratio										
5. Per cent proportion of workers										
6. Per cent proportion of Agricultural workers										
7. Per cent proportion of forest area										
8. Per cent proportion of net cultivated area										
9. Per cent proportion of Irrigated area										
10. Per cent proportion of Electrified villages										
11. Road length per 100 sq.km.										

12. No. of Doctors per 10,000 population	1.00									
13. Hospital Beds per 10,000 population	.38897	1.00								
14. % proportion of canal irrigated area	-.32508	-.11451	1.00							
15. % proportion of double cropped area	.65809	.24198	-.15308	1.00						
16. % proportion of cropped area	.06043	.06186	.07784	.37982	1.00					
17. % proportion of Pasture land	-.00059	-.02203	-.16912	.13057	.00458	1.00				
18. % proportion of population in seconddary activities	.02065	.41461	-.27803	-.01852	-.00934	.02965	1.00			
19. % proportion of follow lands	.58113	.22200	-.46525	.43229	-.09483	.26829	.09308	1.00		
20. % proportion of commercial crops	.58582	.02642	-.54978	.48398	.15594	.14064	-.10457	.64295	1.00	
21. % prop. vill. of H.School	.10042	.07102	-.30811	.27603	.33465	.24918	.16805	.25413	.43379	1.00

TABLE 6.3 : Iteration Cycle : Variances

Iteration Cycle	Variances
0	.203465
1	.407680
2	.456308
3	.457068
4	.457097
5	.457098
6	.457098
7	.457098
8	.457098
9	.457098
10	.457098

TABLE 6.4 : Check of Communalities

Variable	Original	Final	Difference
1	.80846	.80846	.00000
2	.90580	.90580	.00000
3.	.73803	.73803	.00000
4.	.90667	.90667	–.00000
5.	.74013	.74013	–.00000
6.	.81279	.81279	.00000
7.	.82130	.82130	.00000
8.	.91776	.91776	.00000
9.	.87498	.87498	–.00000
10.	.91794	.91794	.00000
11.	.65282	.65282	.00000
12.	.88585	.88585	.00000
13.	.82740	.82740	.00000
14.	.83446	.83446	.00000
15.	.80379	.80479	.00000
16.	.89294	.89294	.00000
17.	.77937	.77937	.00000
18.	.76142	.76142	.00000
19.	.74713	.74713	.00000
20.	.79752	.79752	.00000
21.	.60279	.60279	.00000

TABLE 6.5 : Factor Matrix

	Variables	Fact I	Fact II	Fact III	Fact IV	Fact V	Fact VI	Fact VII
		1	2	3	4	5	6	7
1.	Density of Population	.833	.239	.143	.119	-.001	-.012	.145
2.	% prop. of .S.C. population	-.237	.308	.076	-.271	-.412	.475	-.528
3.	% crop. of S.T. population	-.696	.336	-.206	.000	-.105	.290	-.035
4.	Literacy ratio	.719	.563	.065	.170	.031	-.098	.167
5.	% prop. of workers	-.495	-.627	-.105	.136	.172	.182	.088
6.	% prop. of agri. workers	.261	-.740	-.169	-.029	-.179	.161	.329
7.	% poop. of forest area	-.758	.439	-.108	.195	-.001	.029	.044
8.	% prop of net cult. area	.376	-.527	.657	-.008	.218	.002	-.133
9.	% prop. of irrigated area	.061	-.134	-.000	.831	.292	.146	-.235
10.	% electrified villages	.084	-.226	.605	-.328	.212	.480	.329
11.	Road length per 100 sq. km.	.615	.081	.416	.065	-.119	-.117	-.247
12.	No. of Doctors per 10,000 pop.	.457	-.081	-.682	..250	-.362	.0	-.082
13.	Hospital beds per 10,000 pop.	.468	.316	-.060	.374	-.120	.487	.334

(Contd.)

TABLE 6.5 : *Contd.*

Variables	1	2	3	4	5	6	7
14. % prop. of canal irrigated area	.414	-.143	.435	.643	.101	-.142	-.085
15. % prop. of double cropped area	.460	-.449	-.457	.368	-.089	.182	-.064
16. % prop. of cropped area	.495	-.601	.440	.054	-.243	.084	-.149
17. % prop. of pasture land	.155	-.063	-.167	-.174	.679	.279	-.390
18. % prop. of population in secondary activities	.545	606	.222	.018	.050	.203	.035
19. % prop. of fallow lands	.488	-.067	-.877	-.076	.191	.043	.029
20. % prop. of area under commercial crops	.469	-.286	-.628	-.289	-.079	-.069	-.076
21. % prop. of villages with high schools	-.624	-.178	.030	-.269	.321	-.048	-.044

Factor III (Electrification Dimension)

The third important factor has an eigen value of 3.09 individually explaining 14.7 per cent of the total variance. However, it brings out the positive association between two variables only, as is evident from the following table.

Factor III

	Indicator	Rotated Factor Loading
1.	Proportion of electrified villages	+ 0.51
2.	Proportion of canal irrigated area	+ 0.42
3.	Doctors/population ratio	- 0.91
4.	Proportion of double cropped area	- 0.77
5.	Proportion of commercial crops	- 0.74
6.	Proportion of fallow lands	- 0.72

The association of electrified villages and canal irrigation area could perhaps be explained since increased agricultural prosperity brings in a facility like electricity. However, the negative associations are difficult to be accounted far beyond its statistical factuality in the absence of detailed probing.

Factor IV (Irrigation Dimension)

The fourth factor with an eigen value of 1.92581 explains a little less than ten per cent of total variance. It high-lights the association between irrigation Variables like irrigation area and canal irrigated area. The factor indi-cates negative factor loadings with regard to villages with high schools, electrified villages and commercial crops. It could perhaps be interpreted to indicate that provision of irrigation facilities have not yet reached its optimal level so as to enable the farmers to raise commercial crops. It could also mean that due to problem of water logging during monsoon season, only non-comme-rcial crops like paddy are cultivated.

Factor IV

	Indicator	Rotated Factor Loading
1.	Irrigated area	+ 0.90
2.	Canal irrigated area	+ 0.73
3.	Workers proportion	+ 0.23
4.	Forest area	+ 0.20
5.	Commercial cropped area	- 0.38
6.	% Villages with high schools	- 0.22
7.	% Electrified villages	- 0.21

The rest of the Factors *viz.*, V, VI and VII together explain just 16 per cent of total variance. Therefore, they have not been discussed in detail.

Spatial Pattern of Levels of Development and Backwardness

An attempt was made in the preceding section of the present chapter, to amalgamate selected indicators of development and backwardness by applying Principal Component Analysis technique. The seven factors and the significant associations that emerge within each of the factors explain the relative importance of the indicators in the context of measuring levels of development and backwardness in the command area as a whole. This itself may not be sufficient if the objective is to investigate the spatial spread of the factors. Factors scores for each of the units of study viz., Mandals and for each of the factors have, therefore, been worked out to analyse the spatial patterning of various dimensions of development and backwardness (Table 6.7 & 6.8). Levels of development and backwardness are computed by adopting the following procedural criteria:

1. Factor scores are computed for each of the seven factors and for each of the forty-six Mandals. Appendix 6.2
2. Cartographic representation of Factor scores for the first four Factors is made to get an idea of the spatial spread of different dimensions of development.

TABLE 6.6 : **Rotated Factor Matrix**

	Variables	Factor I	Factor II	Factor III	Factor IV	Factor V	Factor VI	Factor VII
		1	2	3	4	5	6	7
1.	Density of Population	. 66	.710	-.227	-.016	.021	-.218	.261
2.	Per cent proportion of S.C. Population	-.086	.056	.091	-.132	.012	.932	-.006
3.	Per cent proportion of S.T. Population	-.674	-.232	.169	.033	-.105	.427	.066
4.	Literacy ratio	.064	.890	.164	-.002	-.038	-.230	. 66
5.	Per cent proportion of workers	-.036	-.808	.032	.235	.089	-.087	.117
6.	% prop. of agri. workers	.161	-.794	-. 81	.078	-.189	.134	.249
7.	Per cent proportion of forest area	-.757	-.131	.316	.201	-.212	.200	-.066
8.	Per cent proportion of net cultivated area	.927	-.052	.185	.077	-.099	.043	.050
9.	Per cent proportion of irrigated area	.032	.014	.162	.900	.152	-.078	.080
10.	Per cent electrified villages	.400	-.123	. 10	-.218	.230	-.022	.616
11.	Road length per 100 sq. km.	.584	.542	.012	.081	-.003	.049	-.084
12.	Number of Doctors per 10,000 population	.010	.116	-.918	.048	-.148	.057	.015

(Contd.)

TABLE 6.6 : *Contd.*

	Variables	1	2	3	4	5	6	7
13.	Hospital beds per 10,000 Population	-.036	.468	-.298	.185	-.126	.014	.683
14.	Per cent proportion of canal irrigated area	.032	-.199	.425	.737	-.220	-.082	-.112
15.	Per cent proportion of double cropped area	.272	-.125	-.773	.274	.066	-.092	.166
16.	Per cent proportion of cropped area	.923	-.076	-.082	.104	-.055	.045	.106
17.	Per cent proportion of pasture land	-.018	-.002	-.079	.069	.875	.044	-.006
18.	% prop. of population of secondary activities	.047	.805	.046	-.039	.104	.081	.298
19.	Per cent proportion of fallow land	-.081	.111	-.721	-.177	.345	-.231	.054
20.	% prop. of area under commercial crops	151	-.058	-.747	-.380	.202	-.104	-.122
21.	% prop. of villages with high schools	.435	.231	.172	-.223	.452	-.272	.021

TABLE 6.7 : **Factor Scores and Levels of Development**

(Factor–wise I - VII)

Sl. No.	Fac. I	F. Scores	Fac. II	F. Scores	Fac. III	F. Score	Fact. IV	F. Score
	1	2	3	4	5	6	7	8
1.	LD	2.0369	MD	15.7567	MB	6.9983	LB	.4258
2.	LD	1.1943	LD	1.2828	MB	3.8625	LB	.8368
3.	MD	4.9476	MD	2.6415	LD	1.3566	LD	1.1263
4.	MD	2.4067	MD	4.8318	LB	1.0777	LB	1.9691
5.	MD	2.1303	MD	2.5867	LD	.5961	MB	2.4421
6.	MB	-.3060	LD	1.9997	LB	.4409	MB	3.3646
7.	LD	.2546	MD	2.5636	LB	.2469	MB	3.5741
8.	LD	1.1068	LD	2.1542	LB	.6207	MB	3.7466
9.	LD	1.1263	LD	1.3563	LB	.7086	MB	3.8852
10.	MD	4.6043	MD	6.5189	LB	.4698	LD	.8396
11.	MB	1.7527	MB	3.5994	LB	.3122	LD	.6954
12.	LD	.3746	LB	2.406		1.9881	LD	.2469
13.	LB	-4.0574	MB	4.0338	LD	.6272	LB	.6899

(Contd.)

TABLE 6.7 : *Contd.*

Sl. No.	1	2	3	4	5	6	7	8
14.	LD	4898	LB	1.3169	MD	2.3026	LB	1.3321
15.	MD	4.0324	LB	2.5932	LB	1.2103	MD	2.1988
16.	MD	3.6468	MB	3.3646	MD	2.2859	MD	.4355
17.	MB	1.3260	LD	.5579	LB	.9778	MD	1.7280
18.	MD	2.3702	LD	1.9185	MB	2.5671	LD	1.1653
19.	MB	1.8674	MB	3.6461	LD	.5420	LB	.3406
20.	MB	1.2756	MB	4.0764	LD	1.0463	LB	2.6471
21.	MB	2.0484	MB	5.4597	MD	2.4739	MD	2.6693
22.	MD	3.1989	MD	2.4876	MD	2.2167	LD	1.2354
23.	LD	.4495	LD	.7777	LD	1.4546	LB	.1657
24.	MD	5.6466	LD	1.5167	MB	.2377	LB	.2017
25.	LD	1.2760	MD	10. 348	LD	1.7032	LB	1.7971
26.	MB	1.5730	MD	.5594	MD	3.5522	LB	1.7444
27.	MD	3.4141	MD	3.9143	MB	2.3792	LB	2.4955
28.	LD	1.7937	LB	2.5683	LD	1.7086	LD	1.6392
29.	LD	1.4914	LB	2.1548	LD	1.9328	LD	1.6734
30.	MD	4.5096	LB	.2395	LD	1.2005	LD	.7321

31.	LB	4.0994	LD	1.9652	MD	2.0738	LD	.7724
32.	LB	9.0531	LB	.9391	MD	4.7713	LB	.6304
33.	LB	3.6152	MD	3.5092	LB	.9916	LB	1.5490
34.	MB	.0862	LB	3.2182	MD	3.1641	LB	.0405
35.	LB	8.5453	MB	4.4183	LD	.4913	MB	2.6148
36.	LB	9.1761	LB	4.1458	MD	4.018	MD	2.6362
37.	LB	8.7838	MB	4.27	MD	4.0086	MD	2.4965
38.	LB	7.0409	MB	4.0233	MD	4.6011	MD	2.6674
39.	LD	.1876	MB	3.7503	MD	3.9517	MD	2.1349
40.	LD	1.653	LB	.3412	MD	.2855	MD	1.7666
41.	MB	.8487	MB	5.3359	MB	1.2951	MB	3.1510
42.	MD	4.4767	MB	4.0146	MB	1.4365	LB	1.9930
43.	MD	4.6134	LB	1.4720	MB	4.9308	LB	.4936
44.	MD	5.7042	LB	.3008	MB	3.3806	LB	1.4353
45.	MB	1.0577	LB	.0039	MB	13.0583	LB	.1349
46.	MB	.6227	LD	2.3972	MB	11.1285	LB	.5419

Contd.

TABLE 6.7 : *Contd.*

Sl. No.	Fac. V	F. Score	Fac. VI	F. Score	Fac. VII	F. Score
	9	10	11	12	13	14
1.	MD	1.7374	MB	2.5804	MD	5.3981
2.	MD	8.1413	MB	1.2007	LD	.4579
3.	LD	.7063	MB	1.0188	LB	.1002
4.	LD	.3239	MB	.6151	MD	1.2150
5.	MD	.9864	MB	6164	LB	.1214
6.	LD	.5683	LB	.5157	LB	.3492
7.	MD	1.3742	LB	.9619	MB	.3036
8.	MD	1.1121	LB	.5986	MB	.5411
9.	MD	. 545	LB	.6018	MB	.7138
10.	MB	1.2076	LB	.5423	MD	1.9630
11.	LB	.0299	LB	.3826	LD	.4007
12.	MB	1.1183	LD	.3280	LB	.1782
13.	LB	.1417	LB	.2169	LD	.7256

14.	LB	.2768	LB	.0521	LB	.2326
15.	LD	.5793	MB	.9292	LD	.3017
16.	MB	.9732	LD	.0463	LD	.0846
17.	LD	.1576	MB	.8679	MD	1.8334
18.	MD	1.5732	MB	1.9004	MD	1.8018
19.	LD	.3177	LB	.3662	LD	.0983
20.	LD	.1496	LB	.5706	LD	.6181
21.	MB	1.1238	LB	.0530	LB	.6485
22.	LB	.5955	LB	.2555	LB	.2402
23.	LD	.3064	LD	.0567	LD	.1866
24.	MD	1.4678	MB	.2132	LB	.0474
25.	LD	.3387	LB	.0096	MD	1.0068
26.	LB	.4263	LB	5.3462	MD	.7033
27.	LD	.7060	LB	.3146		2.8357
28.	LB	.3525	LD	.1759	LB	.4419
29.	LB	.1268	LD	.1889	LB	.2353
30.	LD	.5176	MB	.7792	LB	.1638

(Contd.)

TABLE 6.7 : *Concld.*

Sl. No.	9	10	11	12	13	14
31.	MB	1.9427	MD	1.3098	MB	2.0576
32.	MB	2.0459	MD	7.4614	LB	.2586
33.	LD	1.3389	MB	.3164		2.0196
34.	LB	.2635	LD	.5192	MB	.7621
35.	LB	.1986	MD	1.5230	MB	1.8864
36.	MB	2.3161	MD	1.4857	MB	1.8599
37.	MB	1.8231	MD	1.7593	MB	1.6656
38.	MB	2.1890	MD	1.5411	MB	1.8193
39.	MD	1.8220	MB	.7292	MD	.5536
40.	MB	1.2675	LB	.0087	LB	.2583
41.	LB	.6740	LB	.3679	LD	.7212
42.	LD	.3496	MB	.6056	MD	.7281
43.	LB	.1254	MB	.8920	MD	2.0041
44.	MD	.7018	MB	1.3855	MB	1.0568
45.	MD	.8175	MB	1.3190	MB	1.6924
46.	MD	.9024	MB	1.4970	MB	2.1323

TABLE 6.8 : **Sriramsagar Command Area : Levels of Development and Backwardness by Seven Factors**

Sl. No.	Mandal	Factor I	Factor II	Factor III	Factor IV	Factor V	Factor VI	Factor VII
		1	2	3	4	5	6	7
1.	Karimnagar	LD	MD	.B	LB	MD	MB	MD
2.	Marakandur	LD	LD	MB	LB	MD	MB	LD
3.	Choppadandi	MD	MD	LD	LD	LD	MB	LB
4.	Huzurabad	MD	D	LB	LB	LD	MB	MD
5.	Jammikunta	MD	MD	LD	MB	MD	MB	LB
6.	Veenavanka	MB	LD	LB	MB	LD	LB	LB
7.	Kamalpur	LD	MD	LB	MB	D	LB	MB
8.	Elkathurthi	LD	LD	LB	MB	MD	LB	MB
9.	Keshavapatnam	LD	LD	LB	MB	MD	LB	MB
10.	Jagtial	MD	MD	LB	LD	MB	LB	MD
11.	Raikal	MB	MB	LB	LD	LB	LB	LD
12.	Dharmapuri	LD	LB		LD	MB	LD	LB
13.	Sarangapur	LB	MB	LD	LB	LB	LB	LD
14.	Mallial	LD	LB	MD	LB	LB	LB	LB

(Contd.)

TABLE 6.8 : *Concld.*

Sl. No.	Mandal	1	2	3	4	5	6	7
15.	Pagadapalli	MD	LB	LB	MD	LD	MB	LD
16.	Gollapalli	MD	MB	MD	MD	MB	LD	LD
17.	Metpalli	MB	LD	LB	MD	LD	MB	MD
18.	Kortala	MD	LD	MB	LD	MD	MB	MD
19.	Medipalli	MB	MB	LD	LB	LD	LB	LD
20.	Ibrahimpatnam	MB	MB	LD	LB	LD	LB	LD
21.	Mallapur	MB	MB	MD	MD	MB	LB	LB
22.	Peddapalli	MD	MD	MD	LD	LB	LB	LB
23.	Srirampur	LD	LD	LD	LB	LD	LD	LD
24.	Odela	MD	LD	MB	LB	MD	MB	LB
25.	Ramagundam	LD	MD	LD	LB	LD	LB	MD
26.	Kamanpur	MB	MD	MD	LB	LB	LB	MD
27.	Sultanabad	MD	MD	MB	LB	LD	LB	-
28.	Velgaturr	LD	LB	LD	LD	LB	LD	LB
29.	Dharmaram	LD	LB	LD	LD	LB	LD	LB
30.	Julapalli	MD	LB	LD	LD	LD	MB	LB
31.	Manthani	LB	LD	MD	LD	MB	MD	MB
32.	Mutharam	LB	LB	MD	LB	MB	MD	LB
33.	Nirmal	LB	MD	LB	LB	LD	MB	

34.	Laxmanchanda	MB	LB	MD	LB	LB	LD	MB
35.	Mamda	LB	MB	LD	MB	LB	MD	MB
36.	Khanapur	LB	LB	MD	MD	MB	MD	MB
37.	Kaddam	LB	MB	MD	MD	MB	MD	MB
38.	Jannaram	LB	MB	MD	MD	MB	MD	MB
39.	Dandepalli	LD	MB	MD	MD	MD	MB	MD
40.	Laxettipet	LD	LB	MD	MD	MB	LB	LB
41.	Chityal	MB	MB	MB	MB	LB	LB	LD
42.	Mogullapalli	MD	MB	M3	LB	LD	MB	MD
43.	Parkala	MD	LB	MB	LB	LB	MB	MD
44.	hampet	MD	LB	MB	LB	MD	B	MB
45.	Atmakur	MB	LB	MB	LB	MD	MB	MB
46.	Hasanparthi	MB	LD	MB	LB	M	MB	MB

Note : MD - More developed; LD - Less developed; MB - More Backward; LB - Less Backward

3. Mandals are grouped into four categories of levels of development and backwardness in each of the factoral dimensions (Table 6.7). All the Mandals with positive scores are considered as relatively developed and the mandals with negative values are identified as backward areas. The whole set of observations or scores for each of the seven factors is divided into four categories as given below, with median value as the dividing line in the case of both positive and negative score distributions.

	Criteria	Level of Development	Code
I.	Mandals with factor score of more than median value of the positive observations in the particular set of factor scores	More Developed	MD
II.	Mandals with factor score of less than median value of the positive observations in a particular set of factor scores	Less Developed	LD
III.	Mandals with factors score of Less than median value of the negative observations in a particular set of factor scores	Less Backward	LB
IV.	Mandals with factor score of more than median value of the negative observations in a particular set of factors scores	More Backward	MB

4. Final identification of composite picture of backwardness is made by examining the More Developed (MD), Less Developed (LB), More Backward (MB) and Less Backward (LB) values of each of the factors for each of the mandals. Spatial patterning of the levels of development and backwardness are thus identified. Following is the category-wise

list of the Mandals:

Category I : More Developed Mandals

1. Karimnagar
2. Jagtial
3. Gollapalli
4. Peddapalli
5. Srirampur
6. Ramagundam
7. Dandepalli
8. Koratla
9. Huzurabad.

Category II : Less Developed Mandals

1. Manakondur
2. Choppadandi
3. Jammikunta
4. Dharmapuri
5. Pagadapalli
6. Metpalli
7. Sultanabad
8. Velgatur
9. Julapalli
10. Manthani
11. Khanapur
12. Jannaram
13. Lakshettipet
14. Shampet

Category III : Less Backward Mandals

1. Kamalapur
2. Elkaturthi
3. Keshavapatnam
4. Mallial
5. Medipalli
6. Ibrahimpatnam
7. Mallapur
8. Velgatur
9. Dharmaram
10. Julapalli
11. Manthani
12. Khanapur

13. Jannaram
14. Lakshettipet
15. Shampet

Category IV : More Backward Mandals

1. Hasanparthi
2. Atmakur
3. Mogulapalli
4. Chityal
5. Mamda
6. Veenavanka
7. Raikal
8. Sarangapur

A closer examination of categorisation of Mandals by their levels and types of development and backwardness unravels some interesting patterns. Nearly all the urban areas, all the former taluk headquarters of the command area are ensconced in the category of more developed mandals. Same is the case with the industrial areas in the north-eastern part of command area. The areas which were provided with irrigation facility in the initial stage of command area development are also well developed. The mandals which are categorised as more developed have some characteristic features in common, specially, with regard to uniformly high density of population, higher proportion of non-agricultural workers/agricultural workers and relatively better endowment with infrastructural facilities like schools, hospitals, other services, road etc. These are the areas where developmental activity has been thriving on its own, mainly because of the factor of historical inertia. Mandals forming part of more developed category do not require as much attention and investment from governmental agencies, since, they have the necessary inherent capability for development because of internal dynamism.

The second category of 'less developed' areas, includes such Mandals which have either recently achieved developmental break-through or, on the verge of achieving it in the near future. They possess great potential of attaining higher level of development provided the necessary incentives are given. Flooding the area with financial assistance is certainly not a desirable one; instead, emphasis should be laid on making optimum utilisation of the already available resources. A great economic future awaits this area.

The category of 'less backward' areas fall into two types: the areas in the irrigated sector and the areas in the unirrigated sector of the command area. It is quite understandable if the areas in the presently unirrigated sector exhibit the characteristic features of 'less backward' areas. The areas in the western part which received irrigation facility as far back as early seventies still remain backward need to be probed to examine the factors that were responsible for its backwardness. It can only be interpreted to mean that mere provision of irrigation facilities without proper back-up facilities would not automatically lead to the development of the area. It may, however, be mentioned here that some of the western parts of command areas are thickly forested and the terrain is not suitable for any other economic activity except forestry.

The 'more backward' areas of the command area lie at the eastern periphery and in the western part where physiographic factors impede agricultural development. These are the areas with a relatively low densities of population and absence, or, near absence of necessary infrastructural inputs that are necessary for developmental activity. It may not be possible to do much to bring up the physiographically inhospitable area. However, a lot of attention both in financial terms and administrative terms, needs to be given to the 'more backward' areas lying in the east. These are the areas which are on the verge receiving irrigation facilities in the near future. As such, proper preparation in physical, economic and social directions is necessary to reap full benefits from the provision of water. These areas are at present having a completely different agricultural system which differs distinctly from the system which will be ushered in, in the near future. The local populace should be made aware of intensive agricultural techniques.

REFERENCES

Ashok Mitra (1961) : *Levels of Regional Development in India* : Census of India 1961, Para 1A(1).

Drewnowski Jan, (1970) *Studies in the Measurement of levels (1967) of Living and Welfare*, Report no.703 UNRISD

Fruchter, B., *Introduction to Factor Analysis.*

Iyengar, N.S. and P. Sudarshan (1980): *On A Method of Classifying regions from Multi-Variate data.*

Nanjappa M.B., *Backward Areas in Mysore state: A Study in Development.*

Granchar, M.C. *et al.* (1970) : *Contents and Measurement of Socio-Economic Development.*

Rao, R.R.M., and Simhadri, S., (1985) : "A Method for Measuring Agricultural Efficiency" in *Readings in Agricultural Resource Development* by K.V. Reddy and R.R.M. Rao.

Rao, R.R.M. (1981) : *Growth of Cities,* Inter-India Publications, New Delhi.

Van de Geer (1971) : *Introduction to Multivariate Analysis for Social Sciences.*

SEVEN
Development Strategy

An investigation into the characteristic features of population, agricultural and settlement systems and the existing levels of development and backwardness of the study area was made in the preceding chapters. It is necessary to restate here briefly the main findings, in order to get a comprehensive and panoramic picture of the various attributes. This should enable us to examine the scene much more meaningfully and methodically in suggesting a broad framework of developmental strategy for command areas in general and Sriramsagar Command Area in particular. The present chapter contains two sections: a summarisation of the major findings of the study and formulation of a strategy for command area development.

SUMMARISATION OF THE MAJOR FINDINGS OF THE STUDY

Population Aspects

An analysis of population characteristic features is important not only in the context of over-all development but also for the reason that the man power and its type and quality constitutes a major input in agricultural activity. On set of irrigation facilities and the resultant requirements need evaluation of the potential human resources. Sriramsagar Command Area as a whole registered lesser growth rate in its population during the last two decades, than the other areas in its vicinity. In the absence of detailed field investigations, it is difficult to explain the fall in the growth rate of population. Actually, it should have been otherwise since the provision of irrigation facilities and consequent spurt in developmental activities would have normally been expected to lead for an increase in population by in-migration. However, a micro-level study revealed that the areas which have been benefited by irrigation facility for a longer period of time have experienced higher positive variation of density than the areas which have, either, not received the water so far, or have received only recently.

Provision of assured irrigation facilities often leads to a change in the cropping system from dry crop cultivation to intensive wet crop cultivation which requires female labour in many important operations of rice cultivation specially in levelling, transplanting, weeding and harvesting. These activities are done by the female labour only as a matter of tradition and also economic convenience. It can, therefore, be assumed that the sex ratio in command areas of irrigation projects will be relatively in favour of females than in the un-irrigated dry crop areas. However, it does not mean that irrigation is the sole determinant of sex-ratio. It is just one of the many factors causing imbalances in sex composition. There may be a number of other factors which have the same or greater influences in determining sex ratio. Sriramsagar Command Area has more females per thousand male population than Telangana and Andhra Pradesh. Two possible causative factors can be investigated for relatively higher population of females in Sriramsagar Command Area. Does this pattern emerge because of in-migration of female labour from other areas? However, as a matter of tradition, female labour does not migrate without being accompanied by their male counterparts. Secondly, it is possible that since the area is a rice cultivation zone, more of female labour is required than male labour; the men migrate to work places outside the command area, thus leading to relatively higher proportion of females. How this process takes place is a debatable point in the absence of concrete evidence.

A striking aspect of command area's population is a relatively low level of literacy. The low level of literacy could be due to increase in the quantum of employment caused by intensive cultivation of the land. The school going children would have perhaps found working in the fields more compelling than going to school. However, it is difficult to substantiate this point without further in-depth studies. The urban areas and the surrounding regions exhibit relatively higher levels of literacy. This could be due to provision of educational facilities in urban areas. This aspect leads to one very vital question. Does the provision of facilities for education is the only factor leading to higher levels of literacy? Does economic prosperity has no bearing in increasing the educational consciousness of the people?

Sriramsagar Command Area which experienced a decrease, though insignificant, of its active population in the decade prior to the release of water, registered substantial increase in the working population during the period 1971-81. The proportion of non-workers came down in ample measure during the decade following the release of waters, thus reversing the trend for the period 1961-71. In this respect, the situation is similar to that of Nagarjunasagar Command Area. The relatively lower value of dependency ratio in Sriramsagar Command Area is indicative of the economic buoyancy and the magnitude of available job opportunities. If we examine the intra-command area dependency ratio pattern, it is quite evident that irrigated part of the command area on the whole, has less than one dependant for each worker. Sixty-five per cent of command area, most of which is irrigated, has a dependency ratio of less than 0.95. The areas surrounding towns have a higher value of dependency ratio. The low magnitude of dependency ratio in the irrigated component can be attributed to the booming agricultural activity which, in turn throws open a number of job opportunities, mostly in agricultural sector itself, but sometimes in non-agricultural sector as well. As in the case of other agriculturally important parts of the country, intensive irrigated agriculture requires more labour.

The command area of Sriramsagar as a whole, did not register much increase in the cultivators category and decrease in the agricultural labourers category. This is all the more noteworthy since the benefits of the provision of irrigation facility and development in the agricultural sector used to be considered and still supposed to be considered as an instrument of ushering in an era of economic equality. It clearly indicates that policy of encouraging owner cultivation has not really succeeded as it was planned. It is reflective of the failure of so-called progressive tenancy and land reform laws. However, the presently irrigated area of Sriramsagar

Command Area presents a somewhat different picture as regards the proportion of cultivation in working population. This has been clearly brought out by micro-level study. The contrasting trend in the distribution of cultivators segment of working population is very evident. Most of the irrigated component of the command area has an above average proportion of cultivators. 76.2 per cent of the irrigated area has more than half of its working population classified as owner cultivators, whereas in the unirrigated sector only a meagre 6.7 per cent of the area has more than half of its working population categorised as cultivators. The village sample survey corroborates this fact. It can, therefore, be surmised that the provision of irrigation facilities has led to a noticeable, but not highly significant, change in the agricultural land tenure system. More and more of marginal farmers and small farmers are no more doubling as agricultural labourers, since, the benefits accrued from their small holdings have become sufficient to sustain them.

Another point which emerges out of a cursory look at occupational categories proportions is that the share of primary, activities is still very high. Increased agricultural activity in the comand are as a whole has not led to the emergence of any significant improvement of the secondary sector. The additional production and the resultant monetary benefits did not lead to any appreciable investment in and, enlargement of industries, despite the availability of raw materials for setting up agro-based industries in the command area. It can only be assumed that available surplus income is not being channellised in the area, but is siphoned off to other areas, probably to the large cities.

Agricultural System

The net area irrigated as per cent of net area sown increased tremendously from 1971 to 1981 in all the taluks. Canal irrigated area as per cent total irrigated area has increased in significant proportions. Area sown more than once a year as per cent of net area sown registered general increase except in a few taluks. Total cropped area as per cent of geographical area has not registered any change. But proportion of per capita availability of cropped area has increased slightly whereas, increase is more in the per capita availability of food cropped area. This happened as a direct consequence of the provision of irrigation facility.

The irrigated part of the command area has added more land to its existing arable land during the decade 1971-81. The unirrigated part of the command area has little forest cover except in the taluk of Manthani where the forests occupy more than 60 per cent of the total geographical area. Manthani is one taluk which is yet to receive canal waters. It is to be seen whether, the provision of waters brings down the proportion, of area under forests thereby disturbing the ecological balance.

At present, Manthani is the only taluk having the required area under forest cover. It would not be in the fitness of things if the provision of the canal water is to disturb the already precarious physical environment in the command area.

Provision of dependable irrigational facilities has not led to a very big increase in the net area sown though, in certain parts of the Command Ara, fallow lands and culturable water lands have been converted into cultivable land in the irrigated sector of the Command Area. Sriramsagar Command Area recorded manifold increase in the area under rice cultivation. This is a natural corollary of the provision of water. Rice is the single most important crop accounting for 33 per cent of the cropped area and it is the predominant crop during Kharif season. However, rice does not occupy the first position during Rabi season. Instead, maize is cultivated as a wet-cum-dry crop. In many parts of Telangana, Maize is cultivated mostly as a subsistence crop and is rain-fed. But in Sriramsagar Command Area, this crop is cultivated on commercial basis by applying greater amounts of inputs. Sriramsagar has always been a stronghold for maize cultivation. But after the provision of waters the efficiency with which maize is cultivated has considerably increased. Maize cultivation is on the ascendency both in quantitative and qualitative terms. With the extension of irrigation facilities the importance of maize crop will further go up. In course of time, it may replace jowar as the most important millet crop. There are very good chances of its emerging as a good substitute crop in place of rice specially during those years when the Sriramsagar reservior does not receive its full quota of waters.

There has been a tremendous impact of irrigation on agricultural efficiency after the provision of irrigation. The taluks which were not provided water in 1971 and ranged from low to moderate regions of agricultural efficiency, have now risen to moderate to high efficiency regions after provision of water. The other taluks which are yet to receive water have fallen from high to moderate regions of agricultural efficiency.

Settlements and their Functional Character

An examination of settlements of Sriramsagar Command Area unravels the complex and vibrant processes. Prior to the year 1970, the type of socio-economic evolution of settlements was similar to that of any other backward regions of Telangana and the country at large. After the provision of assured irrigation, gradual but revolutionary processes are in operation. Economy has become vibrant and is exerting its influence in many spheres of human activity. It can be noted that the share of small villages with less than thousand population decreased markedly during the last censal decade. Large sized villages are growing at the cost of small villages.

In a developing economy, rural functions play a vital role. Since the inception of assured irrigation, the command area is emerging as the dominant agricultural region. A strong infrastructural base is necessary in the light of changing socio-economic structure. As the infrastructure and agricultural system in Sriramsagar Command Area are compatible to one another there is need for a close examination of the infrastructural distirbution pattern in the context of integrated development of Command Area.

Social Setting

Indian society has been functioning on caste basis since times immemorial. This was perhaps required in a tradition bound Hindu society for the smooth conduct and distinct demarcation of various economic and social functions that used to be performed by the populace of ancient and medieval India. The continued existence of caste-ridden society in modern and forward looking India can be said to be an anchronism. Despite the stringent economic and constitutional remedial measures taken by the Government in post-independent era, the barriers and limitations imposed by caste structure have not yet been overcome specially in the rural segment of society. This is true in the case of Sriramsagar Command Area also. Broadly speaking, the society in the Command Area is organised mostly on caste lines.

The important castes/caste-groups that are found in Sriramsagar Command Area are Malas, Madigas (Scheduled Castes) Reddies, Tenugus, Yadavas, Gouds, Padmashalis, Artisans etc. The caste composition as such has not altered appreciably during the twenty years period 1961-81. However, the relative importance of some of the castes underwent a sea change specially in the economic sphere. The scheduled castes of Malas and Madigas together accounted for 19.2 per cent of the total households in the sample villages in 1981. A slight decline in the proportion of their population can be noticed during the preceding twenty years period. As was mentioned earlier, the scheduled caste population is mostly engaged as agricultural labour in the fields owned by the people of other castes. Their distribution in the command area is negatively correlated with literate population. The irrigated sector of the study area registered a relatively higher growth of scheduled caste population than the unirrigated sector during the period 1961-1981. This could be due to the intra-command area migration of agricultural labourers from dry unirrigated taluks to the irrigated areas where the scope for working as agricultural labourers is much more. Field investigations revealed that the economic lot of the few scheduled caste people who own their often small land holdings has improved after the provision of irrigation facilities and now they need not supplement their income by duplicating as farm labourers. This has resulted in some shortage of agricultural labour thus causing migration from the eastern part of command area.

The type of the house in which a family lives can be considered as an indicator of that family's economic and social status. Of course, the physical environment of the area also conditions the type and design of the house. The type of building materials used, the size of the structure and its location reflect the social and economic importance of the family. However, it is necessary to be cautious in examining this aspect, since spatial variations exist and the standards of measurements differ from one region to another. To cite an example, a thatched roof house is perhaps the most dominant type of house of even the upper castes in coastal Andhra, whereas it is considered mainly to be the house type of very low income group people in Telangana.

Levels of Development/Backwardness

A major objective of economic and physical planning measures, like commissioning of major irrigation projects, is to reduce regional disparities to attain the cherished goal of social and economic justice. In order to fully understand and gauge the benefits arising out of the implementation of economic and physical plans, it is essential to identify the regions or areas according to their levels of development, typology and dimensions of backwardness.

Measurement of differential levels of development is necessary for strategy formulation for the development of backward areal pockets. Required policies, programmes and plans can be framed based on the specific area characteristics, their requirements and capabilities. The results of principal component analysis of the levels of development in Sriramsagar Command Area indicate that there are seven dimensions which together explain development/backwardness of the area. It may be mentioned here that a maximisation procedure under certain side conditions takes on a very specific and recognizable form, namely, finding eigen vectors.

A closer examination of categorization of Mandals by their levels of development and backwardness unravels some interesting patterns. Nearly all urban areas, all former taluk headquarters of the command area are ensconced in the category of more developed mandals. Same is the case with the industrial areas. The areas which were provided with irrigation facility in the initial stage of command area development are also well developed. The mandals which are categorised as more developed have some characteristic features in common, especially, with regard to uniformly high density of population, higher proportion of either non-agricultural workers or agricultural workers and relatively better endowment with regard to infrastructural facilities like schools, hospitals, services, roads etc. There are the areas where development activity has been thriving on its own mainly because of the factor of historical inertia. Mandals forming part of 'more developed category' need not get as much attention of and investment from governmental agencies since they have the necessary inherent capability for development.

The second category of less developed areas includes such mandals which have either recently achieved developmental break-through, or are on the verge of achieving it in the near future. They possess great potential of achieving higher level of development provided the necessary incentives are given. Flooding the area with financial assistance is certainly not a desirable one; instead, emphasis should be laid on making optimum utilisation of the already available resources. A great economic future awaits this area.

The category of 'less backward' areas fall into two categories: the areas in the irrigated sector and the areas in the unirrigated sector of the command area. It is quite understandable if the areas in the presently unirrigated sector, exhibit the characteristics of less backward areas. However, it raises questions, if the areas in the western part which received irrigation facility as far back as early seventies still remain backward. It can only be interpreted to mean that mere provision of irrigation facilities without proper back-up facilities would not automatically lead to the development of the area.

The 'more backward' areas of the command area are situated in the eastern periphery and in the western part where physiographic factors impede agricultural development. These are the areas with relatively low densities of population and absence, or, near absence of necessary infrastructural inputs that are necessary for developmental activity. It may not be possible to do much to bring up the physiographically inhospitable area. However, a lot of attention, both in financial terms and administrative terms, need to be given to the "more backward" areas lying in the east. These are the areas which are on the verge of receiving irrigation facilities in the near future. As such, proper preparation in physical, economic and social directions is necessary to reap full benefits from the provision of water. The areas are at present having a completely different agricultural system which differs distinctly from the system which will be ushered in the near future. The local populace should be made aware of intensive agricultural techniques.

FORMULATION OF A STRATEGY OF COMMAND AREA DEVELOPMENT

Strategy for Command Area Development

The overall strategy for economic and social development of command area is a composite of sub-strategies relating to agriculture, industry, infrastructural facilities and social aspects. The total efficiency of the strategy and the success of it depends on the complementarities among various sectors and the quality of implementation of a well thoughtout developmental framework. Developmental activity should not necessarily be confined to just one or two aspects of human life, but the entire gamut of it. Economic upliftment can certainly be the main objective of developmental planning. However, care should be taken to see that economic development does not lead to social injustice. It is important to state in this context that development of people living in the command areas should be the only aim. Development of the place should be considered nothing more than a vehicle for attaining development of the people. Development of the place should be considered as only means to an end not an end in itself.

Another feature that should form part of a strategy for command area development is the principle of equity. It is true that not all areas of the command areas are equally capable of development. Some areas like the hilly areas in the north and east cannot be provided with irrigation facilities due to the obstacles posed by terrain. Such areas should not become the areas of neglect and perpetual backwardness. New avenues of developmental strategy should be to narrow down the divide rather than further widen it. The simmering discontent in the tribal population inhabiting forest areas has already been manifested in their covert and covert supports to the extremist elements.

Strategy for Agricultural Development

Any agricultural strategy should have a broader framework of objectives encompassing the entire rural masses. The strategy should aim at contributing to the over-all rate of economic growth and the process of structural transformation of rural economy. The aim should be to achieve maximum farm output at minimum cost by utilising the many agricultural innovations that exist at present. However, care should be taken that only those innovations which are appropriate to the command area should be encouraged for adoption. Agricultural innovations which have the acceptance of local farming community should be encouraged. Achieving a broadly based improvement in rural population should be the criteria. Any agricultural strategy will not be complete without facilitating the process of social modernisation by encouraging widespread attitudinal and behavioural changes in the rural community. The following sub-startegies are suggested for achieving success in the agricultural sector.

1. Selective diffusion of agricultural innovations should be encouraged taking into consideration the capability and efficacy of much innovations in the command area. Education of farmers in the use of sophisticated agricultural implements is a pre-requisite in this process. The nature and pace of mechanisation should be determined.
2. It has been found that there is not even a single farmers' organisation in the command area. In the absence of this, the farmers are not able to organise themselves in putting forward their problems and in seeking their redressal. An organisation of farmers wedded to its social welfare and economic upliftment, it is essential to increase its lobbying capacity. A disorganised farming community becomes an easy prey of the economic exploiters.
3. Credit and marketing facilities in the command area are woefully inadequate. Provision of credit facilities on easy terms will pave the way for increasing the capacity of the farmer to apply more agricultural inputs. At present, the farmers in the command area find it difficult to apply the necessary inputs, mostly because of the limited credit facilities. Of late, the nationalised banks and some farmers' co-operative credit societies

have liberalised their credit lending policies. However, the liberalisation facilities are helping mostly the farmers with medium and large sized land holdings. Marginal and small farmers are yet to come under the umbrella of bank's credit system. This could be due to small and marginal farmer's ignorance of the provision of liberalised lending policies. The policy of credit giving institutions should be to go to the farmers and not *vice versa*, as is the situation at present.

4. The existing land tenure system in command area is superficially a progressive one when compared to the same enacted in other parts of the country. However, doubts are often expressed about its pratical operation. Rich landlords have found many loopholes in the existing system and thus are able to prevent the emergence of a progressive land tenure system. Plugging the loopholes in the present enactments relating to land ceilings and land tenure is very vital if the fruits of the provision of irrigation facilities is to reach the deprived sections of rural society.
5. Rationalisation of modalities of providing irrigation facilities is a pre-requisite for agricultural development of the region. Modalities of water management should be determined keeping in view all the areas in the command area and also all sections of rural society. Field investigations revealed that tail-end farmers are not provided with consistent water supply. The reason given for this state of affairs is that the stored water is not sufficient enough to cater to the needs of the entire command area. If this is the case, the tail-end farmers should be told in advance, of the period of release of waters so that they can plan their cropping pattern compatible with the period of availability of water. They can cultivate short duration crops if the water is made available for shorter periods. Long duration crops can be farmed if the water is assured for longer periods. It is not only the duration of water release that matters, but also the frequency of release of water. Certain crops like rice require watering on continuous basis whereas, certain dry-cum-wet crops like sugar-cane, groundnut and millets require watering at an inter-

val of once or twice a week depending on the type of soil of that area. Introuction of 'warabandi system' in the command area is a step in the right direction. However, the system needs further rationalisation.

6. Evolving the most economic and appropriate cropping pattern in the command area is another important element in farming agricultural strategy. Selection of the type of crop in an area depends on a number of factors *viz.*, farmer's knowledge of cultivation techniques of particular crops, the type of soil, duration of water availability, frequency of water release into field channels, farmer's capacity to apply the necessarily inputs and the marketing facilities. A combination of all these factors and the relative importance of the individual factors determine the type of crop to be grown in the area. However, the cropping pattern that emerges out of such factors need not necessary be the best and the most economic. In such a case, some degree of control by a co-ordinating authority will prove beneficial. A Cropping Pattern Co-ordinating Authority may be set up in the command area. The co-ordinating authority can consist of the representatives of farming community, officials of the representatives of farming community, officials of the existing Command Area Development Authority, Irrigation Development and Other Agencies. Under the present constitutional limitations, it may not be possible to make it obligatory for farmers to grow only the assigned crops. However, the desired cropping pattern in Command Area can be rationalised by taking recourse to a number of indirect methods like, differential rates of land revenue or irrigation cess for different types of crops, imposition of levies, offering of incentives and disincentives respectively, to those farmers who follow the framed policy and those do not adhere to it.

Strategy for Social Development

The objective of any developmental activity should be the betterment of the society as a whole and the people who constitute that society. Provision of physical facilities and economic infrastructure for development becomes meaningless unless the social implications of such planning measures are not rationally considered and their social benefits are not evaluated. It is essential to aim at simultaneous implementation of physical, economic and social developmental schemes. Provision of irrigation facilities is primarily meant for increasing the agricultural productivity levels in the command area in order to uplift the people living in it, by making them available the newly accrued economic benefits of such a provision. However, the final objective should be to pave the way for social development of the command area. It is imperative on the part of planning authorities to oversee that physical and economic developmental schemes result simultaneously in improving the social well-being of the people. There is a possibility of economic benefits being cornered by a few sections of the society in the absence of simultaneous and concurrent social planning. Such an act deprives the benefits to the rest of the society and paves the way for that segment of the society becoming vulnerable and susceptible to all the ill-effects and consequences of social inequalities and social friction.

Sriramsagar Command Area, being part of a country which is wedded to democracy and socialistic pattern of society, should aim at having a society free from social inequalities. It was evident from the discussion in the preceding chapters that Sriramsagar Command Area still continues to have a caste-ridden society. It is true that other areas in the country are also burdened with this. Social friction is bound to cause more harm in an area of great economic vibrancy, than in other economically not so vibrant ones. This is because of the existence of many possibilities for the already rich to corner nearly all the newly accrued benefits at the cost of poorer sections of the society. It is perhaps pertinent to state that the seeds for Naxalite movement were sown in the command area much before it spread to other parts of Telangana. The extremist movement had its repercussions in the form of driving away the rich landlords to urban areas which itself did not bring in social justice and equality. It succeeded in a limited manner in bringing awarness among the rural masses of their rights. Though it would not be correct to state that command area's society does not have caste and class distinction, it can safely be stated that the intensity of caste distinctions is on the declined with more and more deprived sections of the society being able to go up the social and economic ladder.

The social and economic development process initiated in the command area is bound to succeed only if steps are taken for educational development of the people inhabiting the area. Education should not only be considered as direct investment with the gains accruing internally in the form of increased earnings to the educated person, but also as an investment in the functioning of economic and social system at large. These latter gains accrue externally not only to those in whom the educational input is invested, but also to other members of the command area. However, investment in educational sector is characterised by a substantially longer gestation period than that of other types of investments. But this factor should not deter the planners to shy away from allocating the necessary funds, since, such a step would certainly aggravate the situation and hamper the process of trickling down the benefits of economic development for the target groups of population. Education of the masses can be used as instrument to usher in social equality, to prevent friction and to disseminate hi-tech knowledge concerning agriculture, industry etc.

A characteristic feature of Sriramsagar Command Area is the fact that the income generated from the economic affluence of the area being siphoned off into urban centres, leading to an alarming growth of urban centres. Emergence of strong urban centres heralded a new era of social life and acquiring urban life styles. Cinema theatres have mushroormed, buzzing trade has taken roots, and new political power centres have emerged. These factors together led to the emergence of a new class of people. The emerging structure may lead to erosion of some of the traditional relations but may cause establishment of a new social order. The extent of social change, its intensity, spread and consequences should be assessed and a strategy should be worked out in this direction.

Strategy for Industrial Development

Scope for economic development through agricultural sector in the command area is at present enormous. However, further development may not be possible in the future at the present rate, once possibilities for increasing agricultural production reached the saturation point. It is, therefore, essential to diversify the economic base of the command area so that relatively low pace of development in agricultural sector in future is compensated by initiating development in the related sectors. New avenues which generate income, provide employment and contribute to the gross product of the command areas should be found. One such avenue is industrial sector. Encouragement of industries serves twin purposes of keeping the pace of economic development and absorbing the raw materials provided by agricultural sector. Because of assured irrigation facilities and suitable physical and climatic factors, the agricultural sector in the command area at present is and will be in future able to provide raw materials for a number of allied agro-based industries. Maize, cotton, spices and condiments which require sufficient but not abundant water, are suitable for cultivation in the command area. As has already been mentioned earlier in this volume, maize though once a subsistence food crop, has now acquired the colour of commercial crop due to two factors. The first factor relates to the change in food habits of local people who associate eating rice as a sign of relative affluence. The second factor concerns the cost-benefit aspect of maize cultivation. Maize crop can be grown with minimum employment of labour, less water and not much of inputs. Maize is bound to become a major commercial crop in the area provided the demand for the commodity increases. That is why, there exists a strong case for a strategy for establishing maize based industries like poultry and starch factories. Setting up poultry farms can be done at two levels, one at the level of marginal and small farmers who can manage them as subsidary occupation; two, at a larger level where large sized corporate poultry industries can be set up. These huge poultry farms can get their feed requirements from the command area and market their products in metropolitan cities.

The eastern part of Sriramsagar Command Area is richly endowed with coal. A number of coal-based units, like thermal power generating units are already in existence. Steps may be taken to have a much broader base of coal based industries and ancillaries in the command area.

There also exists a great scope for establishing a number of service industries which can cater to the needs of command area community. The purchasing power of the people is bound to rise as they become more and more affluent.

Strategy for Development of Infrastructure

Development of infrastructural facilities is necessary for any meaningful economic and social development. Without adequte infrastructural facilities, the process of improving economic conditions of the people will naturally take a snail's pace and sometimes may even halt progress. Fortunately provision of infrastructural facilities in Sriramsagar Command Area has kept its pace with the rate of economic development. A very high proportion of villages are electrified. Agricultural and veterinary service centres are found. However, the medical facilities are grossly inadequate, so also the facilities for education. There is a need to strengthen facilities in these sectors.

Appendices

APPENDIX 3.1 : **Sriramsagar Command Area : Agricultural Efficiency - 1971**

Sl. No.	Taluk	Land Use Variables							
		ASM a yr. as % of NAS	SS	WSS	T. Crop as % of Geog. area	SS	WSS	Per captia availability of crop area	SS
		1	2	3	4	5	6	7	8
1.	Karimnagar	0.6	42	46	43.2	52	55	0.51	39
2.	Metpalli	4.8	44	48	46.1	54	57	0.55	42
3.	Jagtial	3.9	43		41.0	50	52	0.62	47
4.	Peddapalli	11.0	47	52	57.5	63	66	0.73	54
5.	Manthani	1.4	42	46	21.2	34	36	0.95	69
6.	Huzurabad	23.2	53	58	48.6	56	59	0.55	42
7.	Lakshettipet	15.0	49	54	29.2	41	43	0.79	58
8.	Khanapur	1.7	47	52	21.5	35	37	0.65	49
9.	Nirmal	7.9	45	49	37.7	47	49	0.79	58
10.	Paikal	39.6	62	68	48.2	55	58	0.76	56
11.	Warangal	63.3	74	81	59.0	64	67	0.47	37
	Mean	16.6	-	-	41.2	-	-	0.67	-
	S.D.	19.1	-	-	12.8	-	-	0.14	-

Sl. No.	Taluk	W S S	Per capita avail. FCA	SS	W S S	Culti. waste as % Geog. A.	SS	W S S
		9	10	11	12	13	14	15
1.	Karimnagar	41	0.37	37	398	1.8	48	50
2.	Metpalli	44	0.44	42	44	0.2	41	43
3.	Jagtial	49	0.51	47	49	0.8	45	47
4.	Peddapalli	57	0.64	56	59	3.9	34	36
5.	Manthani	72	0.78	66	69	0.8	45	47
6.	Huzurabad	44	0.40	39	41	4.7	29	30
7.	Lakshettipet	61	0.72	62	65	0.2	41	43
8.	Khanapur	51	0.56	51	54	0.3	42	44
9.	Nirmal	61	0.56	51	54	2.5	43	45
10.	Parkal	59	0.71	61	64	1.2	48	50
11.	Warangal	39	0.38	38	40	0.0	60	63
	Mean	-	0.55	-	-	1.5	-	-
	S.D.	-	0.14	-	-	2.5	-	-

(Contd.)

APPENDIX 3.1 : *Contd.*

Sl. No.	Taluk	Irrigated Variables								
		NTA as % NAS	SS	W S S	CIA as % TIA	SS	W S S	Mean annual rain	SS	W S S
		16	17	18	19	20	21	22	23	24
1.	Karimnagar	30.1	61	67	8.0	47	54	879	34	36
2.	Metpalli	25.2	55	60	3.4	49	56	948	54	57
3,	Jagtial	14.1	44	48	0.5	44	51	944	53	56
4.	Peddapalli	21.5	51	56	0.0	44	50	946	54	57
5.	Manthani	16.5	46	51	3.8	45	52	950	55	58
6.	Huzurabad	35.8	67	74	2.2	45	52	886	36	38
7.	Lakshettipet	15.3	45	49	23.3	54	62	986	66	69
8.	Khanapur	15.3	45	49	73.7	75	86	979	64	67
9.	Nirmal	7.6	37	41	0.0	44	50	965	59	62
10.	Parkal	31.6	62	68	6.0	46	53	921	46	48
11.	Warangal	8.3	37	41	2.7	45	52	940	52	58
	Mean	20.1	-	-	14.8	-	-	933	-	-
	S.D.	9.4	-	-	23.1	-	-	34	-	-

Sl. No.	Taluk	Yield Variabled					
		A. under HYR as % T. Rice	SS	W S S	Area under HYM as % of Total size	SS	W S S
		25	26	27	28	29	30
1.	Karimnagar	19.6	56	62	27.9	68	75
2.	Metpalli	1.6	37	41	8.0	46	51
3.	Jagtial	23.4	60	66	10.7	49	54
4.	Peddapalli	26.5	63	69	8.3	46	52
5.	Manthani	1.5	37	41	12.3	51	56
6.	Huzurabad	12.2	48	53	25.0	65	71
7.	Lakshettipet	12.4	43	47	2.4	40	44
8.	Khanapur	7.1	43	47	2.4	40	44
9.	Nirmal	3.2	39	43	1.9	40	44
10.	Parkal	17.9	54	59	6.0	44	48
11.	Warangal	26.7	63	69	19.8	59	65
	Mean	13.8	-	-	11.3	-	-
	S.D	9.6	-	-	9.1	-	-

(Contd.)

APPENDIX 3.1 : *Concld.*

Sl.No.	Taluk	Agricultural Innovations						Demographic Variables					
		Prop./ P.Till Tractor	Standard	Weighted Standard Score	Electrified Vill. as % Total Vill.	Standard Score	Weighted Standard Score	% Literacy	SS	WSS	Owner Cultivators	SS	WSS
		31	32	33	34	35	36	37	38	39	40	41	42
1.	Karimnagar	35480	57	60	51.1	62	65	20.7	63	66	95.6	55	58
2.	Metpalli	24675	50	52	8.3	37	69	12.2	44	46	100.0	59	62
3.	Jagtial	21829	49	51	14.8	41	43	13.8	37	49	68.1	57	60
4.	Peddapalli	44437	63	66	32.4	51	54	14.3	48	50	55.6	23	24
5.	Manthani	4766	38	40	7.1	37	39	11.0	51	43	87.6	49	51
6.	Huzurabad	13156	43	45	43.7	63	66	14.5	49	51	96.0	56	59
7.	Lakshettipet	13446	43	45	36.4	53	56	15.6	51	54	79.3	42	44
8.	Khanapur	8279	40	42	46.8	59	62	11.6	42	44	89.6	51	54
9.	Nirmal	57382	71	75	26.0	48	50	15.0	50	52	91.0	52	55
10.	Parkal	23716	50	52	12.9	40	42	11.0	41	43	88.6	49	51
11.	Warangal	14780	44	46	44.5	58	61	25.5	74	78	95.6	55	58
12.	Mean	23813	-	-	30.4	-	-	15.0	-	-	88.8	-	-
	S.D.	16123	-	-	17.5	-	-	4.4	-	-	12.4	-	-

Computed from the Data provided by Bureau of Economics and Statistics.

APPENDIX 3.2 : **Sriramsagar Command Area : Agricultural Efficiency - 1981**

Sl.No.	Taluk	Land Use Variables						
		A.S.M. Q.Y. as % of N.A.S	Standard Score	Weighed Standard Score	T.C.A. as % of G.A.	Standard Score	W S S	Per capita Availability of C.A.
1	2	3	4	5	6	7	8	9
1.	Metpalli	17.9	50	55	58.9	55	58	0.72
2.	Jagtial	19.4	52	57	48.4	48	50	0.71
3.	Mallial	25.3	58	64	54.9	52	55	0.77
4.	Sultanabad	35.3	69	78	70.8	62	65	0.85
5.	Peddapalli	11.1	43	47	57.5	54	57	0.75
6.	Manthani	10.3	42	46	34.0	40	42	0.96
7.	Gangadhara	21.1	53	58	59.5	55	58	0.61
8.	Karimnagar	21.4	53	58	55.7	53	56	0.59
9.	Huzurabad	14.4	46	51	69.0	61	64	0.66
10.	Bhimadevarapalli	28.1	61	67	30.5	38	40	0.33
11.	Chityal	17.3	49	54	42.5	45	47	0.78
12.	Parkal	36.3	70	77	75.6	65	68	0.75
13.	Warangal	13.0	45	50	78.6	67	70	0.64
14.	Nirmal	5.9	37	41	44.4	46	48	0.92
15.	Khanapur	18.6	51	56	27.7	36	38	0.68
16.	Lakshettipet	4.6	36	40	28.5	36	38	0.78
	Mean	18.0	-	-	51.0	-	-	0.72
	S.D	9.3	-	-	16.5	-	-	0.14

Contd.)

APPENDIX 3.2 : *Contd.*

Sl. No.	Taluk	Land Use Variables: Standard Score	W S S	P.C.A. of F.C.A.	Standard Score	W S S	Culturable waste as % OF G.A.	Standard Score	W S S
1	2	10	11	12	13	14	15	16	17
1.	Metpalli	50	53	0.68	51	53	0.22	59	62
2.	Jagtial	49	51	0.70	53	56	1.2	53	56
3.	Mallial	54	57	0.75	56	59	0.3	59	52
4.	Sultanabad	59	62	0.82	61	64	3.3	40	42
5.	Peddapalli	52	55	0.73	55	58	6.5	20	21
6.	Manthani	67	70	0.93	68	71	1.1	54	57
7.	Gangadhara	42	44	0.58	45	47	2.4	46	48
8.	Karimnagar	41	43	0.51	40	42	2.7	44	46
9.	Huzurabad	46	48	0.58	45	47	2.0	48	50
10.	Bhimadevarapalli	26	27	0.27	24	25	2.1	47	49
11.	Chityal	54	57	0.72	54	57	0.2	59	62
12.	Parkal	52	55	0.68	51	54	0.2	59	62
13.	Warangal	46	48	0.51	40	42	0.07	66	69
14.	Nirmal	64	67	0.66	50	53	1.9	45	47
15.	Khanapur	47	49	0.63	48	50	0.4	58	61
16.	Lakshettipet	54	57	0.78	58	70	2.4	46	48
	Mean	-	-	0.66	-	-	1.7	-	-
	S.D	-	-	0.15	-	-	1.6	-	-

Sl.No.	Taluk	Irrigation Variables								
		Net. area irri.as % of NAS	Standard Score	Weighted Standard Score	Canal Irri.area as % of total irrigated area	Standard Score	Weighted Standard Score	Mean annual rainfall in mm	Standard Score	Weighted Standard Score
1.	Metpalli	42.4	71	78	36.6	57	66	948	54	57
2.	Jagtial	36.1	61	67	37.6	58	67	944	53	56
3.	Mallial	24.8	41	45	35.9	57	66	944	53	56
4.	Sultanabad	27.9	47	52	28.6	54	62	946	54	57
5.	Peddapalli	32.0	57	58	7.8	44	51	946	54	57
6.	Manthani	27.9	47	52	0.0	42	48	950	55	58
7.	Gangadhara	29.7	50	55	6.7	45	52	886	36	38
8.	Karimnagar	36.0	60	66	2.7	43	49	879	34	36
9.	Huzurabad	28.6	48	53	0.3	43	49	886	36	38
10.	Bhimadevarapalli	25.6	43	47	0.05	42	48	886	36	36
11.	Chityal	33.7	56	62	0.0	42	48	921	46	48
12.	Parkal	31.4	52	57	7.8	44	51	921	46	48
13.	Warangal	33.2	56	62	0.0	42	48	940	52	55
14.	Nirmal	19.5	32	35	18.1	50	58	965	59	62
15.	Khanapur	36.4	61	67	72.2	72	83	979	64	67
16.	Lakshettipet	26.4	44	48	69.7	71	82	988	56	69
	Mean	30.0	-	-	19.0	-	-	933	-	-
	S.D.	.7	-	-	24.2	-	-	34	-	-

(Contd.)

APPENDIX 3.2 : *Contd.*

		Yield Variables					
Sl.No.	Taluk	Yield of Rice per Ha.	Standard Score	Weighted Standard Score	Yield of Maize Per Ha.	Standard Score	Weighted Standard Score
1	2	3	4	5	6	7	8
1.	Metpalli	2706	57	60	2229	42	44
2.	Jagtial	2487	52	55	3705	54	57
3.	Mallial	2788	59	62	2715	46	48
4.	Sultanabad	2554	54	57	3210	50	53
5.	Peddapalli	2281	48	50	4365	59	62
6.	Manthani	3631	77	81	3033	47	49
7.	Gangadhara	2110	44	46	2134	41	43
8.	Karimnagar	1995	42	44	2420	43	45
9.	Huzurabad	2124	45	47	2490	43	45
10.	Bhimadevarapalli	2526	53	56	1698	37	39
11.	Chityal	2617	55	58	2143	41	43
12.	Parkal	2249	47	49	3802	55	58
13.	Warangal	1905	40	42	2608	45	47
14.	Nirmal	1550	32	34	6205	74	78
15.	Khanapur	2209	46	48	3909	56	59
16.	Lakshettipet	2265	48	50	5174	66	69
	Mean	2374	-	-	3240	-	-
	S.D.	463	-	-	1226	-	-

Sl.No.	Taluk	Yield Variables					
		Area under HYR as % of TRA	Standard Score	Weighted Standard Score	Area under HY Maize as % of TMA	Standard Score	Weighted Standard Score
1	2	9	10	11	12	13	14
1.	Metpalli	5.6	31	33	72.9	76	80
2.	Jagtial	43.9	50	53	63.5	57	60
3.	Mallial	42.2	48	50	34.6	56	59
4.	Sultanabad	45.3	51	54	15.6	47	49
5.	Peddapalli	60.6	58	70	4.3	41	43
6.	Manthani	3.5	30	32	0.6	39	41
7.	Gangadhara	49.2	52	55	36.7	58	61
8.	Karimnagar	35.4	46	48	16.9	47	49
9.	Huzurabad	50.4	53	56	13.8	46	48
10.	Bhimadevarapalli	52.1	54	57	13.8	46	48
11.	Chityal	65.7	61	64	28.2	53	56
12.	Parkal	34.9	45	47	9.2	43	45
13.	Warangal	47.3	52	55	60.7	42	44
14.	Nirmal	41.9	49	51	14.4	46	48
15.	Khanapur	85.8	70	73	46.8	62	65
16.	Lakshettipet	47.6	52	55	2.2	40	42
	Mean	44.3	–	–	22.1	–	–
	S.D.	19.9	–	–	19.4	–	–

(Contd.)

APPENDIX 3.2 : *Concld.*

Sl.No.	Taluk	Agricultural Innovations					
		Prop./ PT Tract.	Standard Score	W S S	Elec.Vill. as % of Total Vill.	Standard Score	Weighted Standard Score
1.	Metpalli		58	61	81.2	47	49
2.	Jagtial		64	67	59.5	32	34
3.	Mallial		62	65	61.5	34	36
4.	Sultanabad		61	64	100.0	59	62
5.	Peddapalli		60	63	84.6	49	51
6.	Manthani		48	50	57.7	48	50
7.	Gangadhara		46	48	100.0	59	62
8.	Karimnagar		54	57	96.7	57	60
9.	Huzurabad		47	49	98.2	58	61
10.	B.D. Palli		45	47	97.3	58	61
11.	Chityal		41	43	92.3	54	57
12.	Parkal		44	46	97.7	58	61
13.	Warangal		56	59	94.4	55	58
14.	Nirmal		58	61	93.4	54	57
15.	Khanapur		62	65	83.7	48	50
16.	Lakshettipet		62	65	76.4	43	45
	Mean		-	-	86	-	-
	S.D.		-	-	15	-	-

Sl.No.	Taluk	Demographic Variables					
		% Literacy	Standard Score	Weighted Standard Score	Owner Cultiva-tors	Standard Score	Weighted Standard Score
1.	Metpalli	17.6	46	48	54843	72	76
2.	Jagtial	21.0	51	54	29853	51	54
3.	Mallial	14.5	40	42	21963	44	46
4.	Sultanabad	20.1	50	53	19123	42	44
5.	Peddapalli	24.9	58	61	31957	53	56
6.	Manthani	17.5	45	47	11012	35	37
7.	Gangadhara	20.1	50	53	19123	42	44
8.	Karimnagar	38.0	81	85	19275	41	43
9.	Huzurabad	25.4	59	62	27591	49	51
10.	B.D. Palli	19.2	48	50	27899	49	51
11.	Chityal	13.1	38	40	29172	50	52
12.	Parkal	18.9	48	50	39233	59	62
13.	Warangal	20.4	50	53	32539	53	56
14.	Nirmal	19.4	48	50	29840	51	54
15.	Khanapur	15.7	42	44	14342	38	40
16.	Lakshettipet	21.2	52	57	50113	68	71
	Mean	20.2	-	-	28700	-	-
	S.D.	5.7	-	-	11791	-	-

Computed from the data provided by Bureau of Economics and Statistics

APPENDIX 6.1 : **Development Indicators** **(in percentage)**

Sl. No.	Mandal	Dens of Pop.	S.C. Pop.	S.T. Pop.	Literacy	Workers	Agl. Labours	Forest area	Net cultivated area	Irrigated area	Electrified villages
		1	2	3	4	5	6	7	8	9	10
1.	Karimnagar	674	14.0	0.9	46.7	33.4	5.7	-	27.9	55.9	100
2.	Manakondur	253	19.5	1.0	19.1	48.5	15.7	-	38.9	51.7	100
3.	Choppadandi	441	24.5	0.8	19.5	50.8	15.8	-	61.2	57.0	100
4.	Huzurabad	365	19.8	0.6	27.7	40.2	11.8	-	58.4	34.2	100
5.	Jammikunta	293	21.8	0.5	20.3	41.7	13.4	-	52.6	28.8	100
6.	Veenavanka	293	19.7	0.9	18.2	44.7	16.5	-	40.5	15.0	100
7.	Kamalpur	288	18.2	0.6	20.1	42.	14.0	12.7	38.5	18.5	100
8.	Elkathurthi	287	27.0	0.8	22.1	43.2	15.7	7.7	44.2	20.3	100
9.	K.patnam	230	20.5	0.3	20.1	46.3	15.8	-	44.2	12.6	100
10.	Jagtial	472	12.8	0.6	31.2	44.8	15.2	12.4	68.5	33.3	100
11.	Raikal	156	14.9	6.3	14.0	59.1	26.2	15.1	40.8	43.4	100
12.	Dharmapuri	212	22.2	2.5	14.3	53.4	22.2	19.8	57.9	27.5	95.4
13.	Sarangapur	110	19.9	7.2	11.0	55.8	24.5	11.7	28.8	32.7	91.3

14.	Mallial	193	18.6	0.3	14.2	54.8	20.7	0.9	42.1	16.4	100
15.	Pegadapalli	230	13.8	1.9	14.1	56.9	21.4	0.9	59.7	63.6	100
16.	Gollapalli	194	21.9	0.5	12.3	55.2	25.1	4.2	76.0	53.0	100
17.	Metpalli	242	14.6	2.9	22.9	55.0	18.2	18.0	37.9	61.4	100
18.	Koratla	429	11.8	0.7	24.8	59.5	14.8	12.9	45.4	64.7	100
19.	Medipalli	185	17.7	0.7	12.8	60.6	26.5	25.1	35.4	46.8	100
20.	Ib.patnam	214	14.8	2.4	12.2	61.0	28.4	20.0	42.0	72.7	100
21.	Mallapur	179	15.3	3.9	9.0	59.9	30.6	20.8	32.0	64.6	90
22.	Peddapalli	294	15.6	0.5	23.6	41.5	11.9	13.5	60.5	49.7	100
23.	Srirampur	188	17.4	0.6	15.3	50.7	17.0	10.1	43.4	48.9	100
24.	Odela	236	16.0	0.	18.0	50.0	17.5	–	60.5	46.8	00
25.	Ramgundam	460	21.4	1.2	31.8	31.8	3.6	10.2	48.5	31.8	96.7
26.	Kamanpur	163	18.4	2.2	16.3	41.8	9.2	20.5	44.8	26.9	92
27.	Sultanabad	241	16.9	1.0	20.2	41.4	15.6	–	57.5	62.0	100
28.	Velgatur	174	30.0	0.3	13.3	56.4	19.1	1.6	59.2	55.0	95.8
29.	Dharmaramam	185	22.3	5.7	11.4	55.1	16.	2.3	56.8	52.8	100
30.	Julapalli	201	23.2	0.4	19.5	39.2	17.9	–	73.3	48.8	100
31.	Manthani	119	22.2	4.5	22.6	46.2	10.4	50.7	38.5	36.6	36.6

(Contd.)

APPENDIX 6.1 : *Contd.*

Sl. No.	Mandal	1	2	3	4	5	6	7	8	9	10
32.	Mutharam	146	114.8	36.2	10.1	50.7	15.0	57.8	31.0	35.0	94.4
33.	Nirmal	365	11.0	5.0	31.0	42.0	8.0	16.0	34.0	51.0	33.0
34.	L.chanda	178	17.0	6.0	10.0	58.0	11.0	10.0	57.0	35.0	100
35.	Mamda	59	14.0	23.0	11.0	53.0	16.0	40.0	18.0	23.0	83.0
36.	Khanapur	84	15.4	16.0	19.0	49.0	17.0	75.0	15.0	41.0	69.0
37.	Koddam	78	21.0	18.0	13.0	49.0	21.0	66.0	15.0	53.0	82.0
38.	Jannaram	111	26.0	12.0	14.0	4930	23.0	65.0	27.0	50.0	73.0
39.	Dandepalli	253	22.0	9.0	12.0	53.0	27.0	15.0	61.0	48.0	90.0
40.	Lakshettipet	224	20.0	2.0	21.0	43.0	21.0	16.0	60.0	46.0	90.0
41.	Chityal	162	20.0	4.0	15.0	53.0	48.0	15.0	40.0	9.0	100
42.	Mog.palli	261	21.0	0.5	14.0	50.0	38.0	-	69.0	21.0	100
43.	Parkal	327	23.0	1.0	17.0	55.0	34.0	-	60.0	38.0	100
44.	Shampet	315	20.0	2.0	19.0	44.0	32.0	-	64.0	35.0	100
45.	Atmakur	210	22.0	2.0	18.0	49.0	21.0	-	30.0	56.0	23.0
46.	Hasanparth	52	26.0	-	20.0	46.0	18.0	-	30.0	53.0	15.0

Sl. No.	RD length per 100 sq. km.	No. of doct-ors per 10000 pop.	Beds per 10000 popu-lation	Canal irriga-tion	Double cropped area	Cropped area	Past-ures	Pop. in sec-ond-ary act.	Fallow lands	Comm. crops	Vill. High School
1	11	12	13	14	15	16	17	18	19	20	21
1.	80.4	1.6	16.6	3.5	26.2	35.2	5.7	55.7	45.8	13.6	24.1
2.	39.5	0.4	1.2	-	25.8	51.9	4.7	18.8	39.5	19.6	27.7
3.	100.8	-	-	70.0	4.0	63.7	8.9	16.5	13.9	17.6	25.0
4.	32.7	1.0	4.4	-	7.8	63.0	6.4	25.7	16.5	16.3	16.6
5.	46.2	0.2	-	-	8.3	57.0	6.3	17.3	21.1	6.6	26.3
6.	41.9	0.5	-	-	6.0	42.5	5.2	14.2	34.9	12.4	10.5
7.	49.8	0.2	1.3	-	5.2	40.5	5.4	13.6	26.7	16.4	31.2
8.	51.9	-	-	-	6.9	47.2	4.3	15.3	20.0	30.2	23.0
9.	63.2	0.3	-	-	6.0	46.8	7.4	13.3	25.7	26.1	16.6
10.	102.2	1.4	5.4	46.4	18.8	68.5	3.0	34.0	14.0	5.7	4.2
11.	28.9	0.7	1.4	43.3	22.1	40.8	3.0	14.6	31.0	6.6	10.0
12.	38.4	0.7	1.1	50.6	12.5	57.9	3.6	12.7	12.2	4.0	9.1
13.	28.4	-	-	29.3	11.0	33.2	2.8	11.2	41.2	6.5	

(Contd.)

APPENDIX 6.1 : *Concld.*

Sl.No.	11	12	13	14	15	16	17	18	19	20	21
14.	55.7	0.02	1.1	35.4	10.8	46.5	9.1	18.2	5.7	9.3	13.3
15.	43.8	0.3	-	62.8	34.7	80.4	8.5	14.5	13.9	22.2	18.7
16.	53.2	-	-	63.5	21.5	76.0	6.1	10.9	12.3	1.7	4.5
17.	22.1	0.7	7.0	39.6	18.9	45.2	1.2	21.4	27.4	7.1	13.3
18.	25.0	0.3	1.7	20.3	36.5	62.0	1.3	33.7	17.8	18.0	26.7
19.	25.9	-	-	3.6	15.5	42.7	3.6	15.9	21.7	10.7	10.5
20.	24.1	0.26	1.6	43.4	18.7	42.0	2.5	13.3	16.4	4.0	11.1
21.	20.9	-	-	69.2	10.9	46.0	1.3	3.1	14.7	4.0	4.5
22.	76.8	0.4	0.4	47.7	15.9	70.2	3.7	11.6	6.2	4.4	12.5
23.	64.5	0.05	0.9	6.4	12.8	48.9	5.0	10.8	12.8	66.9	11.8
24.	103.3	0.05	-	27.2	21.2	73.4	3.8	14.9	19.4	11.9	36.4
25.	55.8	-	-	5.6	10.0	48.5	3.7	63.5	11.8	2.3	13.3
26.	50.2	-	-	20.6	8.0	48.4	3.6	36.9	11.2	2.0	8.0
27.	75.1	1.5	13.8	50.0	24.6	71.6	5.6	19.7	17.6	10.7	9.5
28.	52.0	-	-	53.2	12.6	66.7	5.3	11.6	18.8	8	4.2
29.	57.4	0.13	1.6	64.0	11.6	63.4	9.3	9.9	16.2	9.9	11.8
30.	52.8	-	0-	58.9	15.4	73.4	5.6	12.4	16.2	11.3	26.6
31.	63.2	0.26	3.1	13.8	7.7	39.3	1.8	20.9	1.7	1.4	6.7

32.	26.0	0.42	3.4	3.4	2.4	31.8	2.7	13.9	1.1	4.5	–
33.	22.0	–	–	42.5	9.0	37.5	5.8	15.7	10.1	–	5.0
34.	26.0	–	–	20.4	16.5	18.5	7.8	5.4	6.4	–	10.5
35.	14.0	–	–	–	4.8	18.8	22.4	5.0	29.7	29.2	3.3
36.	16.0	–	1.0	85.2	16.5	17.6	0.5	7.3	4.5	–	6.6
37.	24.0	–	–	71.1	12.6	17.3	0.8	0.5	10.5	–	5.4
38.	16.0	–	–	85.1	6.8	28.5	.8	7.8	3.9	–	7.7
39.	41.0	–	–	74.9	3.9	65.9	1.1	5.4	8.8	–	3.2
40.	43.0	–	1.0	67.5	18.2	70.6	2.1	9.0	10.30	–	0.0
41.	11.0	1.0	2.0	–	18.9	48.2	7.5	6.3	13.2	18.5	15.6
42.	22.0	–	–	–	32.4	90.9	4.0	6.3	16.1	18.8	10.5
43.	55.0	2.0	5.0	–	36.7	82.6	2.1	8.0	23.3	18.8	3.0
44.	46.0	1.0	2.0	15.0	33.2	85.7	2.2	7.2	16.8	21.8	33.3
45.	33.0	5.0	2.0	–	46.2	45.2	13.3	6.2	46.1	39.3	17.4
46.	57.0	3.0	–	–	1.2	42.3	2.4	14.3	53.1	37.7	11.0

Bibliography

Alain Birou, *et al.*, 1977, "Towards a Redefinition of Development : Essays and Discussions on the Nature of Development in an International Perspective", Pergamon Press, Paris, p. 9.

Alam, S.M. (ed.), 1974, "Planning Atlas of Andhra Pradesh", Govt. of India & A.P., Hyderabad.

Alden, J., and Morgan R., 1974, "Regional Planning : A Comprehensive View", Leonard Hill Books, Bath, p. 71.

Ali Mohammad, 1978, "Studies in Agricultural Geography", Rajesh Publications, New Delhi.

Andhra Pradesh, "Report of the Annual Survey of Industries, 1982-83", BEST.

Beaujeu-Garnier, J., 1976, "Methods and Perspectives in Geography", Longman, London.

Berry, BJL, and Garrison, W.L., 1958, "The functional bases of Central Place Hierarchy", *Economic Geography*, Vol. 34.

Bhat, L.S., *et al.*, 1976, "Micro Level Planning: A Case Study of Karnal Area, Haryana, India", K.B. Publications, New Delhi, pp. 5, 8.

Bhatia, S.S., 1967, "A New Measurement of Agricultural Efficiency in U.P.", *Economic Geography*, Vol. 43, No. 3.

Blaut, M., 1972, "Space and Process in the conceptual Revolution in Geography" (W.K.D Davies ed.), University of London.

Bourne, C.S., *et al.* (ed.), 1984, "Urbanization and Settlement Systems: International Perspectives", Oxford University Press.

Brown, E.H. (ed.), 1980, "Geography Yesterday and Tomorrow", Oxford University Press.

Castells, Manuel, 1977, "Urban Question, A Marxist Approach" Trans by Alansheridan, Edward Arnold, London.

Census of India, 1961, "Andhra Pradesh: General Population Tables", Part-II-A, Govt. of Andhra Pradesh.

Census of India, 1971, "Andhra Pradesh: General Population Tables", Part-II-A, Govt. of Andhra Pradesh.

Census, 1971, 'District Census Hand Book: Karimnagar Village and Town Directory, Part-X-A, B, Govt. of A.P.

Census of India, 1981, "District Census Hand Book: Karimnagar", Govt. of A.P. Part-XIII-A & B, Series-2.

Chakraborty, Satyesh C., 1986, "On Decentralized Regional Planning in India", in Shafi, Md. Mehdi Raza, (ed.), *Spectrum of Modern Geography*, Concept Publishing Company, New Delhi.

Chakraborty, Sukhmoy, 1984, "Power Structure and Agricultural Productivity" in Desai, *Agricultural Productivity in South Asia*, Oxford University Press, Delhi.

Chenery, Hollis B., 1971, "Studies in Development Planning", Harvard University Press, Cambridge.

Chorley, R.J., Hagett, P., ed., 1978, "Socio-Economic Models in Geography", Methuen & Co. Ltd., London.

Christaller, W., 1933, "Central Places in Southern Germany", translated by Baskin, C.W., Prentice-Hall, New Jeresy, 1966.

Cloke, P.J., 1983, "An Introduction of Rural Settlement Planning", Methuen, London.

Coffey, J.W., 1981, "Geography : Towards General Spatial Systems Approach", Methuen, London.

Cox, P.R., 1976, "Demography", Vikas Publication, New Delhi.

Davies, W.K.D., (ed.), 1972, "The Conceptual Revolution in Geography", University of London Press Ltd.

Dickson, R.E., 1976, "Regional concept: The Anglo-American Leaders", Routledge & Kegan Paul, London.

Dubhashi, P.R., 1983, "Grammar of Planning: Concepts and Applications", IIPA, New Delhi, p.15.

Everett, B.S., & Dunn, G., 1983, "Advanced Methods of Exploration and Modelling", Heinmann, London.

Friedman, J., Douglass, M., 1978, "Agropolitan Development: Towards a New Strategy for Regional Planning in Asia", in Lo, F.C. and K. Salih (ed.), *Growth Pole Strategy and Regional Development Policy*, Oxford: Pergamon Press, pp. 163-92.

Friedmann, John and Clyde Weaven, 1979, "Territory and Function: The Evolution of Regional Planning", Edward Arnold, London, pp. 114, 125.

Fruchetet, B., 1967, "Introduction to Factor Analysis", Von Nostrand Comp., Princeton.

Gadgil, D.R., 1967, "District Development Planning".

Ghosh, Dhira, 1981, "Impact of Irrigation on landuse and cropping Intensity" in Noor Md. (ed.), *Perspectives in Agricultural Geography*, Concept, New Delhi, Vol. II.

Gibbs, Jack P., Ed., 1961, "Urban Research Methods", D-Von Nostrand Co., East West Press Pvt. Ltd., New Delhi.

Gilg, A., 1985, "An Introduction to Rural Geography", Edward Arnold, London.

Glasson, John, 1974, "Introduction to Regional Planning Concepts, Theory and Practice".

Goode, W.H., Hatt, K.H., 1952, "Methods in Social Research", McGraw-Hill Koga Kusha Ltd., Tokyo.

Goryachera, A.M. & Lipeyts, Y.G., 1976, "Inter relation of Factors of Growth in the System of Indian Cities", in S.M. Alam & Pokshishavsky, V.V., (ed.), *Urbanization in Developing Countries*, O.V., Hyderabad.

Government of Andhra Pradesh, 1985, "Statistical Abstract 1983-84".

Haggett, Pater, *et al.*, 1977, "Locational Models", Arnold Heinmann London, pp. 60-62, 146.

Hague, Cliff, 1984, "The Development of Planning Thought: A Critical Perspective", Hutchinson, London.

Hansen, Niles M., (ed.), 1978, "Human Settlement Systems: International Perspectives on Structure, change and Public Policy", Ballinger Publishing Company, Cambridge.

Harinarain, 1980, "Physical Maps of Karimnagar District", Survey of India, Dehradun.

Hooja, Rakesh, 1986, "District Planning: Concept, Setting and State Level Applications", Aalekh Publishers, Jaipur.

Hoyle, B.S., (ed.), 1974, "Spatial Aspects of Development", John Wiley, London.

Jackson, P. & Susan, J.S., 1984, "Exploring Social Geography", George Allen & Unwin, London.

Jae-on-Kim, 1982, "Factor Analysis: Statistical Methods and Practical Issues", Sage Univ. Paper, Sage Publ., London.

Jha, Ashok Kumar, 1980, "Socio-economic and Demographic characteristics and their correlates of the settlements in a Tribal Region: A Case Study of Simega Sub-Division of Ranchi District in Bihar", Mimeo, CSRD, JNU, New Delhi.

Johnson, E.A.J., 1970, "Organization of Space in Developing Countries", Cambridge: Harvard University Press.

Johnston, R.J., 1980, "Multivariate Statistical Analysis in Geography", Longman, London.

Kabra, K.N., 1977, "Planning Process in India".

King, L.J., & Goledge, R.G., 1978, "Cities, space and Behaviour: The Elements of Urban Geograpry", Prentice-Hall Inc., Englewood Cliff, p. 138.

Kundu, Amithab, 1977, "Methods of Regionalisation and their applications to the study of urban processes in N.W. India, 1961-71", Mimeo., CSRD - JNU, New Delhi.

Lal, Tarsem, 1986, "District Development Planning: A Case Study of Two Districts", Concept, New Delhi.

Mahesh Chand & V.K. Puri, 1983, "Regional Planning in India: Some Concepts, Methods and Techniques", Allied.

Majid Hussian, 1979, "Agricultural Geography", Inter-India Publications, New Delhi.

McGranahen, D., *et al.*, 1985, "Measurement and Analysis of Socio-economic Development", U.N. Research Institute for Social Development, Geneva.

Mathews, J.A., 1981, "Quantitative and Statistical Approaches in Geography: A Practical Manual", Pergamon Press, Oxford.

Meier, M. Gerald, 1986, "Leading Issues in Economic Development", Oxford University Press, Delhi.

Misra, G.K., *et al.*, 1987, "Black level Planning : With focus on Employment Generation", Rawat Publ., Jaipur.

Misra, R.P., (ed.), 1981, "Humanising Development: Essays on People, Space and Development", Maruzen Asia, Singapore, pp. 245, 247.

Misra, R.P., 1985, "Development issues of our time", Concept Publishing Company, New Delhi, p. 88.

Misra, R.P., (ed.), 1985, "Rural Development: Capitalist and Socialist path", Concept, New Delhi.

Misra, R.P. *et al.*, 1974, "Regional Development Planning in India: A New Strategy".

Misra, R.P. and Sunderam, K.V., 1980, "Multi-Level Planning and Integrated Rural Development in India", Heritage Publishers, New Delhi, p. 54.

Moser, C.A., & Kalton, G., 1979, "Survey Methods in Social Investigation", Heinmann, London.

Noor, Mohammad, 1981, (ed.), "Perspectives in Agricultural Geography", Concept Publishing Company, New Delhi.

Nanjundappa, D.M., 1983, "Area Planning and Rural Development", APH, New Delhi.

Pandey, Rejendra, 1985, "Sociology of Development: Concepts, Theories and Issues", Mittal Publications, Delhi, p. 126.

Peet, Richard, (ed.), 1977, "Radical Geography: Alternative view points on contemporary social issues", Methuen & Co. Ltd., London.

Planning Commission, 1969, "Guidelines for the Formulation of District Plans", Govt. of India, New Delhi.

Planning Commission, 1978, "Operational Guidelines for Block Level Planning", Mimeo, Govt. of India.

Planning Commission, 1984, "Report of the working group on District Planning", Govt. of India, New Delhi, p. 69.

Prakash Rao, V.L.S., 1984, "Backward Areas in Andhra Pradesh: Patterns of Development", Mimeo, CESS, Hyderabad.

Prasad, Mahavir, 1980, "Sectoral shifts in the work force and the organization of space in developed and less developed districts in the Punjab - 1961-71: Preliminary Exploration", Mimeo, CSRD, JNU, New Delhi.

Rajgopal, M.V., (ed.), 1974, "Andhra Pradesh District Gazetteers - Karimnagar", Govt. of A.P., Kurnool.

Ramachandran, H., 1980, "Village clusters and Rural Development, Concept, New Delhi.

Rameshchand, 1979, "Spatial Organization of Social Facilities: A Case Study of Bulandshahr District", Mimeo, CSRD, JNU, New Delhi.

Rao, M.V.S., 1979, "Socio-Economic Indicators for Development Planning", in UNESCO - The use of Socio-Economic Indicators in Developmnent Planning, Paris.

Rao, R. Ram Mohan and Simhadri, S., 1984, "A Method for Measuring Agricultural Efficiency" in K.V. Reddy and Rao, RRM, (ed.), *Reading in Agricultural Resources Development*, Hyderabad Publishers, Hyderabad.

Raza, Moonis, 1981, "Regional Disparities in India: A Preliminary Exploration of the Regional Dimension of Agricultural Development", in Noor Md., *Perspectives in Agricultural Geography*, Concept, New Delhi, Vol. IV.

Rees, P.H., & Wilson Ag, 1977, "Spatial Population Analysis", Edward Arnold, Bristal.

Sartaz Aziz, 1978, "Rural Development: Learning from China", Macmillan Press, London.

Sen, L.K., *et al.*, 1971, "Planning Rural Growth Centres for Integrated Area Development: A Study in Miryalaguda", NIRD, Hyderabad, pp. 22, 81.

Sen, L.K., *et al.*, (ed.), 1972, "Readings on Micro Level Planning and Rural Growth Centres", NICD, Hyderabad.

Sen, L.K., *et al.*, 1975, "Growth Centres in Raichur: An Integrated Area Development Plan for a District in Karnataka", NICD, Hyderabad.

Sharma, A.N., 1981, "Spatial Approach for District Planning: A Case Study of Karnal District", Concept Publication Co., New Delhi.

Sharma, P.N. & Shastri, C., 1984, "Social Planning: Concepts and Techniques", Print House (India), Lucknow, p. 115.

Simhadri, S., 1983, "Spatial Analysis of Agricultural Efficiency, Land utilization and cropping pattern: Sriramsagar Command Area", (M. Phil. Diss.) Mimeo, Dept. of Geography, Osmania University, Hyderabad.

Simhadri, S., Rao, R. Ram Mohan, 1988, "Water Resource as an Instrument of Social Change", in Vidyanath, V., & Rao, R.R.M., (ed.), Development of *India's Resource Base: Patterns, Problems & Prospects*, Gain Publ., New Delhi.

Singh, Jasbir, 1976, "An Agricultural Geography of Haryana", Vishal Publications, Kurukshetra.

Singh, Jasbir & S.S. Dhillon, 1984, "Agricultural Geography", Tata McGraw-Hill Publishing Corp. Ltd., New Delhi.

Singh, Kanchan, 1981, "Settlement Structural and Process of Regional Development: A Case Study of Bundelkhand", Mimeo, CSRD, JNU, New Delhi.

Singh, K.N. & Singh D.N., 1985, "Rural Development in India: Problems Strategies and Approaches, NGSI, BHU, Varanasi.

Singh, K.N., Singh, S.B., Singh, S.B., 1985, Service centres and Development Strategy in Vindyachal Baghelkhand Region: A Spatio-functional approach", *National Geographical Journal of India*, Vol. 31, (Part-2), BHU, Varanasi.

Singh, S.B., 1977, "Rural Settlement Geography: A Case Study of Sultanpur District, New Delhi, UBBP, Gorakhpur.

Sinha, Ajayakumar, 1981, "Growth of Population and its demographic characteristics in five major irrigation Project Areas: Inter regional and sub-regional variations, 1951, 61, 71", Mimeo, CSRD, JNU, New Delhi.

Sinha, N.K.P., & Rawat, D.S., 1985, "Levels of Development and Development Potential of Chamoli District: An Approach in Integrated Area Development of a Mountainous Terrain", in K.N. Singh & D.N. Singh, (ed.), *Rural Development in India: Problems, Strategies and Approaches*, NGSI, BHU, Varanasi.

Sinha, V.P., 1976, "Chota Nagpur Plateau: A Study in Settlement Geography", K.B. Publication, New Delhi.

Smith, Neil, 1984, "Uneven Development: Nature, Capital and the Production of Space", Basil Blackwell, New York.

Soja, W. Edward, 1980, "The Socio-Spatial Dialectic", Annals of the Association of American Geographer, Vol. 70, No. 2, p. 207.

Stohr, W.B., 1981, "Development from Below: The Bottom-Up and Periphery Inward Development Paradigm", in Stohr, W.B. and Taylor, D.R.F., (ed.) Development from above or below? The Dialectics of Regional Planning in Developing Countries, John Wiley and Sons, New York, pp. 40, 45.

Streeton, P., 1981, "Development Perspectives", Macmillan Press Ltd., London.

Subramanyam, M.J., 1986, "Settlement System and Demographic Structure of Nalgonda District", Mimeo, Dept. of Geography, O.U. Hyderabad.

Sundaram, K.V., 1969, "Settlement Pattern for an irrigated colony; An Application of Central Place Theory" in R.P. Misra and others, Regional Planning: Concepts, Techniques, Policies and case studies, Univ. of Mysore, p. 566.

Sundaram, K.V., 1983, "Geography of Under-Development: The Spatial Dynamics of Underdevelopment". Concept Publishing Co., New Delhi.

Taylor, D.R.F., 1979, "Spatial Aspects of the Development Process", in RA Obudho & DRF Taylor, (ed.), *The Spatial Structure of Development: A Study of Kenya*, Westview Press, Colorado, pp. 1, 4.

Tiwari, R.C. and Khan, N.U., 1984, "Spatial Organization of Rural Service Centres in Pratapgarh Dist". *National Geographer*, Vol. XIX, No.2, Dec., 1984.

Toyne, Peter and Newby, P.T., 1977, "Techniques in Human Geography", Macmillan Edn. Ltd., London, p.116.

Tripathy, R.N., *et al.*, 1979, "Development Plan for a Block Ibrahimpatnam", NIRD, Hyderabad.

Tripathy, R.N., *et al.*, 1980, "Block Plan in the District Frame, A Development Plan for Madakasira Block in Ananthpur Dist., A.P.", Mimeo, NIRD, Hyderabad.

Ullman, Edward, 1970, "A Theory of Location of Cities" in Peter Ambrose, *Concepts in Geography 2*, Analytical Human Geography, Longman, London, pp. 135-142.

Ullman, E.L., 1980, "Geography as Spatial Interaction", Univ. of Washington Press, London.

UNAPDI, 1980, "A Glassory of Terms in Regional Planning and Development", Abhinav Publi., New Delhi.

UNESCO, 1979, "The Use of Socio-Economic Indicators in Development Planning", Paris.

UNESCO, 1981, "Methods for Development Planning: Scenaries, Models and Micro-studies", UNESCO Press, Paris.

UNESCO, 1981, "Socio-economic Indicators for Planning: Methodological Aspects and Selected Examples", Socio-economic Studies - 2, Paris.

Unwin, David, 1981, "Introductory Spatial Analysis", Methuen, London.

Van de Geer, J.P., 1971, "Introduction to Meltivariate Analysis for the Social Sciences". W.H. Freeman and Comp., San Fransisco.

Vidyanath, V., 1985, "Agricultural Productivity in Andhra Pradesh", The National Geographical, *Journal of India*, Vol. 31, Pt. 3, p. 204.

Wanmali, S., 1970, "Regional Planning for Social Facilities; An Examination of Central Place Concepts and their Application; A Case Study of Eastern Maharashtra", NIRD, Hyderabad.

Wanmali, S., 1972, "Location of Growth Centres", in L.K. Sen, and Readings in Micro-level Planning Rural Growth Centres, NICD, Hyderabad.

Wanmali, S., 1981, "Periodic Markets and Rural Development in India", B.R. Publishing Corpn., Delhi.

Wanmali, S., 1983, "Service Provision and Rural Development in India: A Study of Miryalaguda Taluka", Research Report (37) International Food Policy Research Institute.

Whyte, R.O., 1982, "The Spatial Geography of Rural Economics", Oxford University Press, Bombay.

Zelinsky, W., 1966, "A Prologue to Population Geography", Prentice-Hall Inc., New Jeresy.

Zutshi, Bupinder, 1976, "Settlement Hierarchy and Economic Structure of Villages in the Kashmir Valley", Mimeo, CSRD, JNU, New Delhi.

Index